AF270835

General Biology Laboratory Manual for Science Majors

BIOL 1406 LAB

Custom Edition for North Lake College

James W. Perry | David Morton | Joy B. Perry

Journi Norman

BIOL 1406 LAB

CENGAGE Learning

Australia • Brazil • Japan • Korea • Mexico • Singapore • Spain • United Kingdom • United States

General Biology Laboratory Manual for Science Majors: BIOL 1406 LAB Custom Edition for North Lake College

Senior Project Development Manager:
Linda deStefano

Market Development Manager:
Heather Kramer

Senior Production/Manufacturing Manager:
Donna M. Brown

Production Editorial Manager:
Kim Fry

Sr. Rights Acquisition Account Manager:
Todd Osborne

LABORATORY MANUAL FOR MAJORS GENERAL BIOLOGY, 1ST EDITION
Perry | Morton | Perry
© 2009 Cengage Learning. All rights reserved.

LABORATORY MANUAL FOR NON-MAJORS BIOLOGY, SIXTH EDITION
Perry | Morton | Perry
© 2013, 2007 Cengage Learning. All rights reserved.

ALL RIGHTS RESERVED. No part of this work covered by the copyright herein may be reproduced, transmitted, stored or used in any form or by any means graphic, electronic, or mechanical, including but not limited to photocopying, recording, scanning, digitizing, taping, Web distribution, information networks, or information storage and retrieval systems, except as permitted under Section 107 or 108 of the 1976 United States Copyright Act, without the prior written permission of the publisher.

For product information and technology assistance, contact us at
Cengage Learning Customer & Sales Support, 1-800-354-9706

For permission to use material from this text or product,
submit all requests online at **cengage.com/permissions**
Further permissions questions can be emailed to
permissionrequest@cengage.com

This book contains select works from existing Cengage Learning resources and was produced by Cengage Learning Custom Solutions for collegiate use. As such, those adopting and/or contributing to this work are responsible for editorial content accuracy, continuity and completeness.

Compilation © 2014 Cengage Learning
ISBN-13: 978-1-305-00783-3

ISBN-10: 1-305-00783-2

Cengage Learning
5191 Natorp Boulevard
Mason, Ohio 45040
USA
Cengage Learning is a leading provider of customized learning solutions with office locations around the globe, including Singapore, the United Kingdom, Australia, Mexico, Brazil, and Japan. Locate your local office at:
international.cengage.com/region.

Cengage Learning products are represented in Canada by Nelson Education, Ltd.
For your lifelong learning solutions, visit **www.cengage.com/custom.**
Visit our corporate website at **www.cengage.com.**

Printed in the United States of America

Brief Contents

Preface

Greetings from the authors! We're happy that you are examining the results of our efforts to assist you and your students. We believe you'll find the information below a valuable introduction to this laboratory manual.

Audience

This manual is designed for students at the college majors level. You'll find that the exercises support any biology text used in a majors course.

Features of This Edition

Inquiry Experiments and the Methods of Science

We realize that the best way to learn science is by doing science. Thus, in this edition, you will find numerous **inquiry-based experiments**. Some are extensions of a preceding activity, while others stand alone. Each follows the cognitive techniques whose foundations are laid in the first exercise, "Scientific Method."

Throughout the manual, we place more emphasis on testing predictions generated from hypotheses. We strongly believe that all biology students—benefit by repeating the logical thought processes of science.

New and Updated Exercises

As always, we strive to provide students with exciting, relevant activities and experiments that allow them to explore some of the rapidly developing areas of biological knowledge. We have added an additional animal diversity exercise and re-ordered and re-grouped some of the microbe, plant and invertebrate phyla.

Updated Taxonomy

An attempt to provide accurate systematic information is like trying to hit a moving target. As new information floods in, our best understanding of taxonomy and systematic relationships sometimes seems to change daily. The taxonomy in our exercises is completely updated and reflects the most widely accepted information at the time of publication.

What's Important to Us

In preparing this lab manual, we paid particular attention to pedagogy, clarity of procedures and terminology, illustrations, and practicality of materials used.

Pedagogy

The exercises are written so the conscientious student can accomplish the objectives of each exercise with minimal input from an instructor. As suggested by the publisher, the procedure sections of the exercises are more detailed and step-by-step than in other manuals. Instructions follow a natural progression of thought so the instructor need not conduct every movement.

We attempted to make each portion of the exercise part of a continuous flow of thought. Thus, we do not wait until the post-lab questions to ask students to record conclusions when it is more appropriate to do so within the body of the procedure. Answers to in-lab questions are to be found in the *Instructor's Manual* as well as online.

Terms required to accomplish objectives are **boldface**. Scientific names and precautionary statements, or those needing emphasis, are *italic*.

The use of scientific names is deemphasized when it is not relevant to understanding the subject. However, these names generally do appear in parentheses because the labels on many prepared microscope slides bear only the scientific name.

Format

Each exercise includes:

1. Objectives: a list of desired outcomes.

2. Introduction: to stimulate student interest, indicate relevance, and provide background.

3. Materials: a list for each portion of the exercise so a student can quickly gather the necessary supplies. Materials are listed "Per student," "Per pair," "Per group," and "Per lab room."

4. Procedures: including safety notes, illustrations of apparatus, figures to be labeled, drawings to be made, tables to record data, graphs to draw, and questions that lead to conclusions. The procedures are listed in easy-to-follow numbered steps.

5. Pre-lab Questions: ten multiple-choice questions that the student should be able to answer after reading the exercise, but prior to entering the laboratory.

6. Post-lab Questions: questions that draw on knowledge gained from doing the exercise and that the student should be able to answer after finishing the exercise. These post-lab questions assess **recall** (preparing students for lab practical assessments), **understanding,** and **application.**

Practical Post-lab Questions

Virtually all courses use laboratory practical examinations. We explain to our students the difference between lecture-type questions, which they need to read and provide an answer based on the written word, and practical-type questions, for which a response depends on observation.

We believe the post-lab questions should draw on the knowledge gained by observation. Consequently, we've incorporated as many illustrations as possible into the post-lab questions. These illustrations typically are similar, but not identical, to those in the procedures. Thus, they assess the student's ability to use knowledge gained during the exercise in a new situation.

Post-lab questions are identified by exercise section. This allows the instructor to easily assign or use only those questions relevant to portions of exercises performed by students. The questions also have been revised and are directly tied to the learning outcomes expected from each exercise.

Flexible Quiz Options

Each exercise has a set of pre-lab questions. We have found through nearly 80 combined years of experience that students left to their own initiative typically come to laboratory unprepared to do the exercise. Few read the exercise beforehand. One solution to this problem is to incorporate some sort of graded pre-exercise activity. At the same time, we recognize that grading a large number of lab papers each week can put an unreasonable burden on instructors. Consequently, we decided on a multiple-choice format, which is easy to grade but still accomplishes the pedagogical goal.

In our own courses, we have students take a pre-lab quiz consisting of the questions in their lab manual in scrambled order to discourage memorization. These scrambled quizzes are reproduced in our *Instructor's Manual.* Our quiz takes about five minutes of lab time and counts as a portion of the lab grade, thus rewarding students for preparation.

Other instructors have told us they use the pre-lab questions to assess learning after the exercise has been completed. We encourage you to be creative with the manual; do what you like best.

Lab Length and Exercise Options

We realize there is wide variation in the amount of time each instructor devotes to laboratory activities. To provide maximum flexibility for the instructor, the procedure portions of the exercises are divided by major headings, and the approximate time it takes to perform each portion indicated. Once the introduction has been studied, portions of the procedures can be deleted or conducted as demonstrations without sacrificing the pedagogy of the exercise as a whole.

It's our experience that if the lecture section covers the topic prior to the lab, students find the exercise much more relevant and understandable. We strive to create this situation in our courses, and thus no time is spent on a lecture-style introduction in the lab itself before the exercise begins. Therefore, we have two to three full hours for real scientific investigation and need to delete very little material to complete most exercises in the time allotted.

Illustrations

Perhaps our illustrations are more noticeable than anything else as you thumb through the manual. We continued our incorporation of a generous number of high-quality color illustrations, including everything a student needs visually to accomplish the objectives of each exercise. While there is no need for students to purchase supplemental publications, students may find the *Photoatlas for Biology*, ISBN 0-534-23556-5, to be a useful reference for this course and other biology courses.

Most illustrations of microscopic specimens are labeled to provide orientation and clarity. A few are unlabeled but are provided with leaders for students to attach labels. In other cases, more can be gained by requiring the student to do simple drawings. Space is included in the manual for these, with boxes for drawings of macroscopic specimens and circles for microscopic specimens.

Materials

Most of the equipment and supplies used in the exercises are readily available from biological and laboratory supply houses. Many others can be collected from nature or purchased in local supermarket, discount or office supply stores. (Our department budgets are not large!) We've attempted to keep instrumentation as simple and inexpensive as possible.

Anyone who lives in a temperate climate knows it may be necessary to adjust the sequence of the laboratory exercises to accommodate seasonal availability of certain materials. However, we provided alternatives, including the use of preserved specimens wherever possible, to avoid this problem.

Instructor's Manual

There is no need to worry, "Where can I get that?" or "How do I prepare this?" Our *Instructor's Manual* includes:

- Material and equipment lists for each exercise
- Procedures to prepare reagents, materials, and equipment
- Scheduling information for materials needing advance preparation
- Approximate quantities of materials needed
- Answers to in-text questions
- Answers for pre-lab questions
- Answers for post-lab questions
- Tear-out sheets of pre-lab questions, in scrambled order from those in the lab manual, for those who wish to duplicate them for quizzes
- Vendors and ordering information for supplies

And in the End . . .

There are very few things in life that are perfect. We don't suppose that this lab manual is one of them. We hope your students will enjoy the exercises. We **know** they will learn from them. Perhaps you and they will find places where rephrasing will make the activity better. Please contact us with your opinions and any ideas you wish to share; encourage your students to do likewise.

James W. Perry
Department of Biological
Sciences
University of Wisconsin
Fox Valley
1478 Midway Road
Menasha, WI 54952-1297
(920) 832-2610
james.perry@uwc.edu

David Morton
Department of Biology
Frostburg State University
Frostburg, MD 21532-1099
(301) 689-4355
dmorton@frostburg.edu

Joy B. Perry
Department of Biological
Sciences
University of Wisconsin
Fox Valley
1478 Midway Road
Menasha, WI 54952-1297
(920) 832-2653
joy.perry@uwc.edu

About the Editor

Mark Manteuffel holds a doctorate in ecology and evolutionary biology from the University of Miami. He is currently a faculty member at Washington University and St. Louis Community College, where he involves students in research learning experiences and teaches courses in biology, ecology, and environmental science. He teaches field courses around the world that focus on the conservation of biodiversity, sustainable development, and ecosystem function and health.

Professor Manteuffel also develops innovative interdisciplinary undergraduate and graduate science programs that cover global education and environmental sustainability. This focus led him to work with colleagues from the University of Missouri, Columbia, converting laboratory investigations into more inquiry-based and student-directed biological investigations. This work is currently ongoing, funded by National Science Foundation grant DEB-0618817, *Connecting Undergraduates to the Enterprise of Science* (CUES).

The inquiry-based lab modules in this edition of the laboratory manual are an extension of Professor Manteuffel's work on the CUES grant. These modules are designed to introduce students to the process and content of science. Students read and critique the lab modules' scientific reports and extend the research reported. The goal of these modules is for students to learn not only content matter, but also the process of science through their own direct experience. They design their own research that will extend the research reported in the modules. The intent is that students will benefit from learning science by doing science, thereby increasing their ability to read and interpret scientific literature.

To the Student

Welcome! You are about to embark on a journey through the cosmos of life. You will learn things about your-self and your surroundings that will broaden and enrich your life. You will have the opportunity to marvel at the microscopic world, to be fascinated by the cellular events occurring in your body at this very moment, and to gain an appreciation for the environment, including the marvelous diversity of the plant and animal world.

We offer a number of suggestions to make your college experience in biology a pleasant one. We have taken the first step toward that goal; we have written a laboratory guide that is user friendly. You will be able to hear the authors speaking as though we were there to share your experience. The authors share a personal belief that the more comfortable we make you feel, the more likely you will share our enthusiasm for biology. One thing we all must realize is that we are citizens of "spaceship Earth." The fate of our spaceship is largely in your hands because you are the decision makers of the future. As has been so aptly stated, "We inherited the earth from our parents and grandparents, but we are only the caretakers for our children and grandchildren."

As caretakers, we need to be informed about the world around us. That's why we enroll in colleges and universities with the hope of gaining a liberal education. In doing so, we establish a basis on which to make educated decisions about the future of the planet. Each exercise in this manual contains a lesson in life that is of a more global nature than the surroundings of your biology laboratory.

To enhance your biology education, take the initiative to give yourself the best possible advantage. Don't miss class. Read your text assignment routinely. And, read the laboratory exercise before you come to the lab.

Each exercise in the manual is organized in the same way:

1. Objectives tell exactly what you should learn from the exercise. If you wish to know what will be on the exam, consult the objectives for each exercise.

2. The Introduction provides background information for the exercise and is intended to stimulate your interest.

3. The Materials list for each portion of the exercise allows you to determine at a glance whether you have all the necessary supplies needed to do the activity.

4. The Procedure for each section, in easy-to-follow step-by-step fashion, describes the activity. Within the procedure, spaces are provided to make required drawings. Questions are posed with space for answers, asking you to draw conclusions about an activity you are engaged in. You'll find a lot of illustrations, most of which are labeled and others which are not but have leaders for you to attach labels. The terms to be used as labels are found in the procedure and in a list accompanying the illustration. We believe it best for you to sometimes make a simple drawing, and have inserted boxes or circles for your sketches. Where appropriate, tables and graphs are present to record your data.

5. Pre-lab questions can be answered easily by simply reading the exercise. They're meant to "set the stage" for the lab period by emphasizing some of the more salient points.

6. Post-lab questions are intended to be answered after the laboratory is completed. Some are straightfor-ward interpretations of what you have done, while others require additional thought and perhaps some research in your textbook. In fact, some have no "right" or "wrong" answer at all!

It is our experience that students are much too reluctant to ask questions for fear of appearing stupid. Remember, there is no such thing as a stupid question. Speak up! Think of yourselves as "basic learners" and your instructors as "advanced learners." Interact and ask questions so that you and your instructors can fur-ther your/their respective educations.

Laboratory Supplies and Procedures

Materials and Supplies Kept in the Lab at All Times

The following materials will always be available in the lab room. Familiarize yourself with their location prior to beginning the exercises.

- Compound light microscopes
- Dissection microscopes
- Glass microscope slides
- Coverslips

- Lens paper
- Tissue wipes
- Plastic 15-cm rulers
- Dissecting needles
- Razor blades
- Assorted glassware-cleaning brushes

- Detergent for washing glassware
- Distilled water
- Hand soap
- Paper towels
- Safety equipment (see separate list)

Laboratory Safety

None of the exercises in this manual are inherently dangerous. Some of the chemicals are corrosive (causing burns to the skin) and others are poisonous if ingested or inhaled in large amounts. Contact with your eyes by otherwise innocuous substances may result in permanent eye injury. **Remember, once your sight is lost, it's probably lost forever.** Locate the following safety items and then study the list of basic safety rules.

1. Eyewash bottle or eye bath

   Should any substance be splashed in your eyes, wash them thoroughly.

2. Fire extinguisher

   Read the directions for use of the fire extinguisher.

3. Fire blanket

   Should someone's clothing catch fire, wrap the blanket around the individual and roll the person on the floor to smother the flames.

4. First-aid kit

   Minor injuries such as small cuts can be treated effectively in the lab. Open the first-aid kit to determine its contents.

5. Safety goggles

   Eye protection should be worn during the more experimental exercises.

Safety Rules

1. Do not eat, drink, or smoke in the laboratory.
2. Wash your hands with soap and warm water before leaving the laboratory.
3. When heating a test tube, point the mouth of the tube away from yourself and other people.
4. Always wear shoes in the laboratory.
5. Keep extra books and clothing in designated places so your work area is as uncluttered as possible.

6. If you have long hair, tie it back when in the laboratory.
7. Read labels carefully before removing substances from a container. Never return a substance to a container.
8. Discard used chemicals and materials into appropriately labeled containers. Certain chemicals should not be washed down the sink; these will be indicated by your instructor.

Caution: Report all accidents and spills to your instructor immediately!

Instructions for Washing Laboratory Glassware

1. Place contents to be discarded in proper waste container as described in exercise.
2. Rinse glassware with tap water.
3. Add a small amount of glassware cleaning detergent.
4. Scrub using an appropriately sized brush.

5. Rinse with tap water until detergent disappears.
6. Rinse three times with distilled water (dH_2O).
7. Allow to dry in inverted position on drying rack (if available).

When glassware is clean, dH_2O sheets off rather than remaining on the surface in droplets.

The Scientific Method

After completing this exercise, you will be able to

1. define *scientific method, mechanist, vitalist, cause and effect, induction, deduction, experimental group, control group, independent variable, dependent variable, controlled variables, correlation, theory, principle, bioassay;*

2. explain the nature of scientific knowledge;

3. describe the basic steps of the scientific method;

4. state the purpose of an experiment;

5. explain the difference between cause and effect and correlation;

6. describe the design of a typical research article in biology;

7. do a bioassay as an example of an experiment.

Introduction

To appreciate biology or, for that matter, any body of scientific knowledge, you need to understand how the scientific method is used to gather that knowledge. We use the scientific method to test the predictions of possible answers to questions about nature in ways that we can duplicate or verify. Answers supported by test results are added to the body of scientific knowledge and contribute to the concepts presented in your textbook and other science books. Although these concepts are as up to date as possible, they are always open to further questions and modifications.

One of the roots of the scientific method can be found in ancient Greek philosophy. The natural philosophy of Aristotle and his colleagues was mechanistic rather than vitalistic. A mechanist believes that only natural forces govern living things, along with the rest of the universe. A vitalist believes that the universe is at least partially governed by supernatural powers. Mechanists look for interrelationships between the structures and functions of living things and the processes that shape them. Their explanations of nature deal in **cause and effect**—the idea that one thing is the result of another thing. (For example, fertilization of an egg initiates the developmental process that forms an adult.) In contrast, vitalists often use purposeful explanations of natural events. (The fertilized egg strives to develop into an adult.) Although statements that ascribe purpose to things often feel comfortable to the writer, try to avoid them when writing lab reports and scientific papers.

Aristotle and his colleagues developed three rules to examine the laws of nature: first, carefully observe some aspect of nature; second, examine these observations as to their similarities and differences; and third, produce a principle or generalization about the aspect of nature being studied. An example is the principle that all mammals nourish their young with milk, which made it difficult to accept two egg-laying mammals found in Australia.

The major defect of natural philosophy was that it accepted the idea of absolute truth. This belief suppressed the testing of principles after they had been formulated. Another example is Aristotle's belief in spontaneous generation, the principle that some life can arise from nonliving things (e.g., maggots from spoiled meat). This belief survived more than 2000 years of controversy before being discredited by Louis Pasteur in 1860. Rejection of the idea of absolute truth coupled with the testing of principles either by experimentation or by further pertinent observation is the essence of the modern scientific method.

1.1   Modern Scientific Method *(About 70 min.)*

Although there is not one universal scientific method, Figure 1-1 illustrates the general process.

MATERIALS

Per lab room:

- blindfold
- plastic beakers with an inside diameter of about 8 cm stuffed with cotton wool
- four or five liquid crystal thermometers

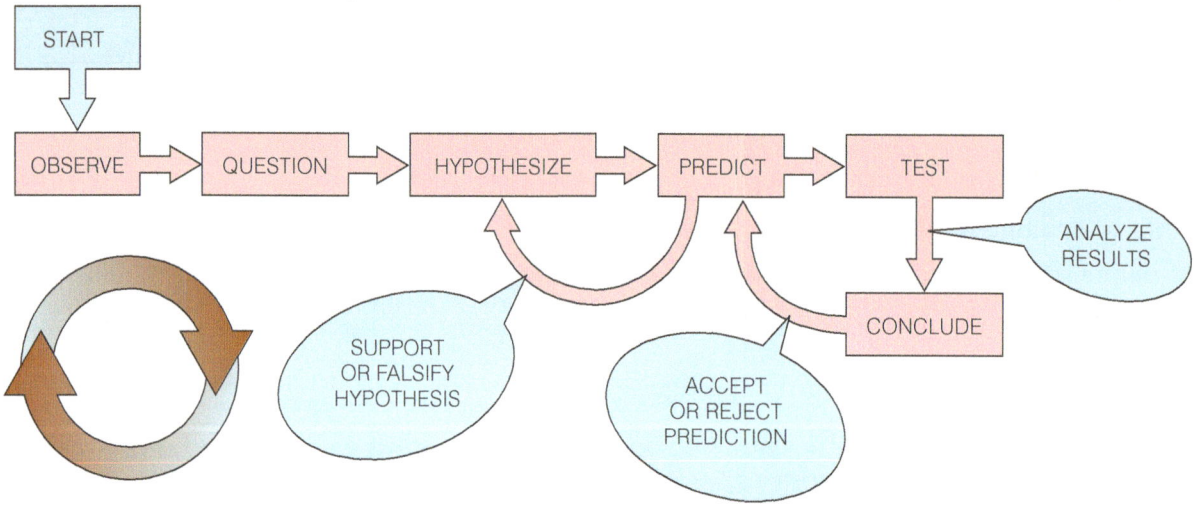

Figure 1-1 The scientific method. Support of the hypothesis usually necessitates further observations, adjustment to the question, and modification of the hypothesis. Once started, the scientific method cycles over and over again, each turn further refining the hypothesis.

PROCEDURE

A. Observation. As with natural philosophy, *we start the scientific method with careful observation*. An investigator may make observations from nature or from the words of other investigators, which are published in books or research articles in scientific journals and are available in the storehouses of human knowledge (e.g., libraries and the Internet). One subject we all have some knowledge of is the human body. The first four rows of Table 1-1 list some observations about the human body. The fifth row is blank so that you can fill in the steps of the scientific method for either the bioassay described at the end of the exercise, another observation about the human body, or anything else you and your instructor wish to investigate.

B. Question. In the second step of the scientific method, *we ask a question* about these observations. The quality of this question will depend on how carefully the observations were made and analyzed. Table 1-1 includes questions raised by the listed observations.

C. Hypothesis. Now *we construct a hypothesis*—that is, we derive by inductive reasoning a possible answer to the question. Induction is a logical process by which all known observations are combined and considered before producing a possible answer. Table 1-1 includes examples of hypotheses.

D. Prediction. Next *we formulate a prediction*—we assume the hypothesis is correct and predict the result of a test that reveals some aspect of it. This is deductive or "if-then" reasoning. Deduction is a logical process by which a prediction is produced from a possible answer to the question asked. Table 1-1 lists a prediction for each hypothesis.

E. Experiment or Pertinent Observations. Now *we perform an experiment or make pertinent observations* to test the prediction.

1. Along with the other members of your lab group, choose one prediction from Table 1-1. Coordinate your choice with the other lab groups so that each group tests a different prediction.

2. In an experiment of classical design, the individuals or items under study are divided into two groups: an **experimental group** that is treated with (or possesses) the independent variable and a **control group** that is not (or does not). Sometimes there is more than one experimental group. Sometimes subjects participate in both groups, experimental and control, and are tested both with and without the treatment.
   In any test there are three kinds of variables. The **independent variable** is the treatment or condition under study. The **dependent variable** is the event or condition that is measured or observed when the results are gathered. The **controlled variables** are all other factors, which the investigator attempts to keep the same for all groups under study.

3. Here are some hints about each of the predictions listed in Table 1-1.
   (a) To test the prediction in row I of Table 1-1, follow these directions:
       (i) With arms dangling at their sides, identify in as many group members as possible a vein segment either on the back of the hands or on the forearms (Figure 1-2).
       (ii) Demonstrate that blood flows in vein segments: Note their collapse due to gravity speeding up blood flow when each subject's arm is raised above the level of the heart.
       (iii) Lower the arm to a position below the heart and stop blood flow through the vein segment by permanently pressing a finger on the swelling farthest from the heart.

TABLE 1-1 Some Observations about the Human Body

Observation	Question	Hypothesis	Prediction
I. Veins containing blood are seen under the skin. Swellings present along the vein are often located where veins join together.[a]	What is the function of the swellings?	Swellings contain one-way valves that allow blood to flow only toward the heart.	If these valves are present, then blood flows only from vein segments farther from the heart to the next segments nearer the heart and never in the opposite direction.
II. People have two ears.	What is the advantage of having two ears?	Two ears allow us to locate the sources of sounds.	If the hypothesis is correct, then blocking hearing in one ear will impair our ability to determine a sound's source.[b]
III. People can hold their breath for only a short period of time.	What factor forces a person to take a breath?	The buildup of carbon dioxide derived from the body's metabolic activity stimulates us to take a breath.	If the hypothesis is correct, then people will hold their breath a shorter time just after exercise compared with when they are at rest.
IV. Normal body temperature is 98.6°F.	Is all of the body at the same 98.6°F temperature?	The skin, or at least some portion of it, is not 98.6°F.	If the hypothesis is correct, then a liquid crystal thermometer will record different temperatures on the forehead, back of the neck, and forearm.
V. _____	_____	_____	_____

© Cengage Learning 2013

[a]The portion of the vein between swellings is called a segment and is illustrated in Figure 1-2.
[b]This is especially the case for high-pitched sounds because higher frequencies travel less easily directly through tissues and bones.

(iv) Use a second finger to squeeze the blood out of the vein segment past the next swelling toward the heart.

(v) Remove the second finger and note whether blood flows back into the vein segment.

(vi) Remove the first finger and note whether blood flows back into the vein segment.

(vii) Repeat v–vi, only reverse the order of the swellings—first stop blood flow through the swelling nearest the heart, then squeeze the blood out of the vein segment in a direction away from the heart past the next swelling.

(b) To test the prediction in row II of Table 1-1, you will need a blindfold and a beaker stuffed with cotton wool to block hearing in one ear. Quantify how well each blindfolded group member can point out the source of a high-pitched sound by estimating the angle (0–180°) between the line from source to subject and the line along which the subject points to identify the source.

(c) To test the prediction in row III, each group member in turn will need to exercise in a safe way. Either run in place or *follow instructions given to you by your instructor*.

(d) To test the prediction in row IV, you will need a liquid crystal thermometer to measure skin temperature in three places for each group member.

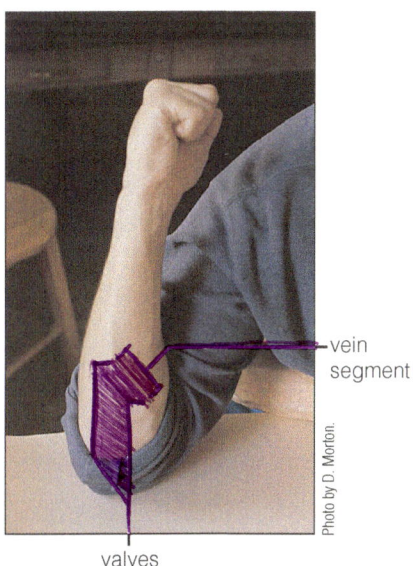

vein segment

valves

Photo by D. Morton.

Figure 1-2 Veins under the skin.

4. In your own words describe your procedure to test your group's prediction and identify its variables below.

_____

_____

_____

Independent variable:

_____

_____

Dependent variable:

_____

_____

Controlled variables:

_____

_____

5. Show the above section to your instructor and get approval to perform your test.

F. Test. After your instructor's approval, perform your procedure and record your results in Table 1-2. You will need to label the column headings for your particular test. You may not need all of the rows and columns to accommodate your data. Describe or present your results in one of the bar charts, etc. in Figure 1-3.

G. Conclusion

1. In the last step of one cycle of the scientific method, *we make a conclusion.* Use the results of the experiment or pertinent observations to evaluate your hypothesis. If your prediction does not occur, it is rejected and your hypothesis or some aspect of it is falsified. If your prediction does occur, you may conditionally accept your prediction, and your hypothesis is supported. However, you can never completely accept or reject any hypothesis; all you can do is state a probability that one is correct or incorrect. To quantify this probability, scientists use a branch of mathematics called *statistical analysis.*

2. Even if the prediction is rejected, this does not necessarily mean that the treatment caused the result. A coincidence or the effect of some unforeseen and thus uncontrolled variable could be causing the result. For this reason, the results of experiments and observations must be *repeatable* by the original investigator and others.

3. Even if the results are repeatable, this does not necessarily mean that the treatment caused the result. *Cause and effect*, especially in biology, is rarely proven in experiments. We can, however, say that the treatment and result are correlated. A correlation is a relationship between the independent and the dependent variables.

   The following example should illustrate these concepts. Severe narrowing of a coronary artery branch reduces blood flow to the heart muscle downstream. This region of heart muscle gets insufficient oxygen and cannot contract and may die, resulting in a heart attack. The initial cause is the narrowing of the artery and the final effect is the heart attack. Perhaps the heart attack victim smoked cigarettes. Smoking cigarettes is one of several factors that make a person more likely to have a heart attack. This is based on a correlation between smoking and heart attacks in the general population but we cannot say for sure that the smoking caused the heart attack.

TABLE 1-2	Generalized Data Sheet		
Subject			
1			
2			
3			
4			
5			
Average			

© Cengage Learning 2013

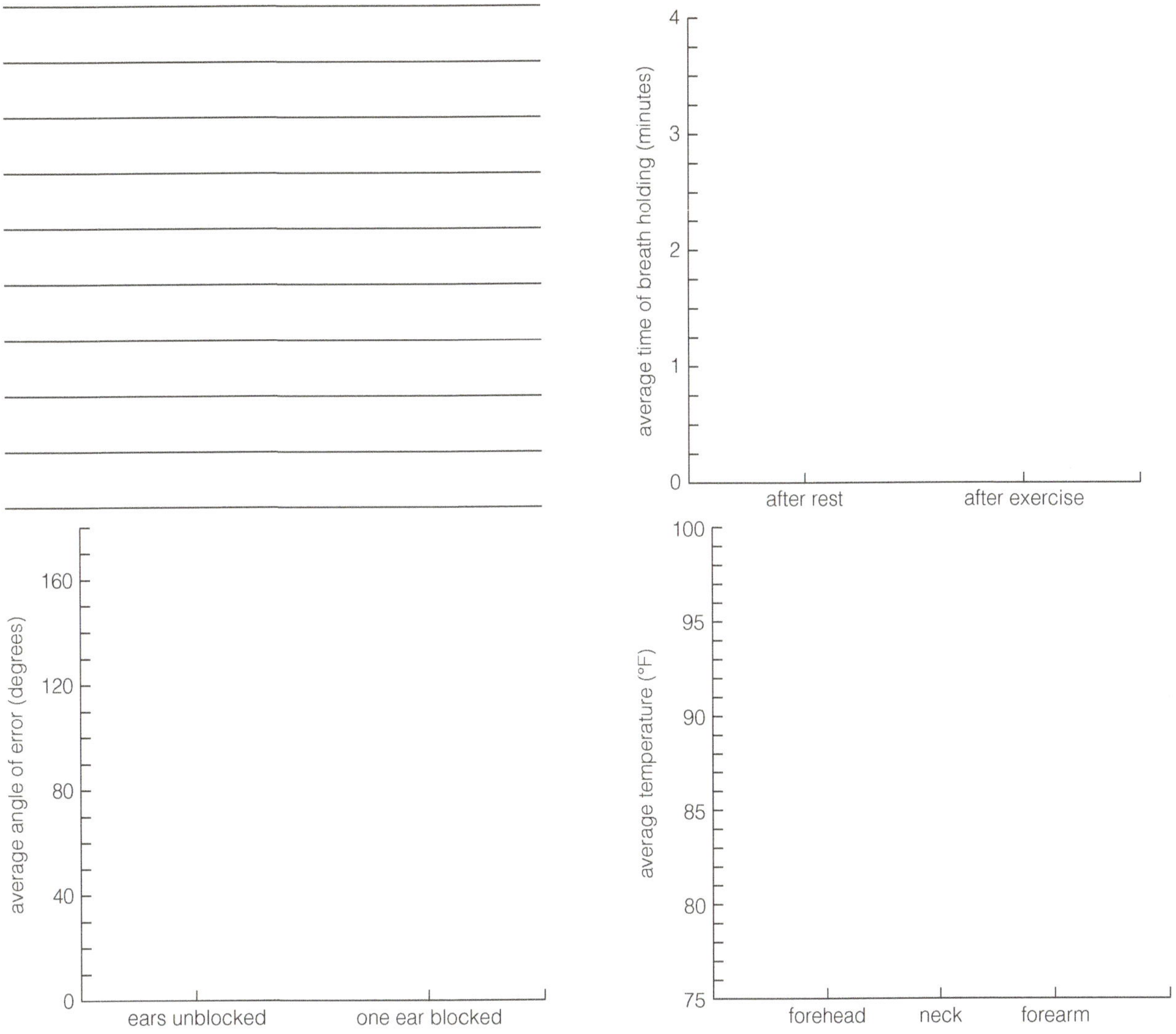

Figure 1-3   Results.

4. Write a likely conclusion for your experiment or pertinent observation. Statistics are not required, but if you know how, apply the correct statistical test before writing your conclusion.

H. Theories and Principles.   When exhaustive experiments and observations consistently support an important hypothesis, it is accepted as a theory. A theory that stands the test of time may be elevated to the status of a principle (Figure 1-4). Theories and principles are always considered when new hypotheses are formulated. However, like hypotheses, theories and principles can be modified or even discarded in the light of new knowledge. Biology, like life itself, is not static but is constantly changing.

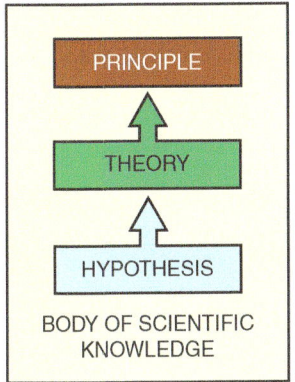

Figure 1-4   Theories and Principles. When exhaustive experiments and observations consistently support an important hypothesis, it is accepted as a **theory.** A theory that stands the test of time may be raised to the status of a **principle.** Theories and principles are always considered when new hypotheses are formulated. However, like hypotheses, theories and principles can be modified or even discarded in the light of new knowledge. Biology, like life itself, is not static but is constantly changing.

The account of one or several related cycles of the scientific method is usually initially reported in depth in a research article published in a scientific journal. The goal of the scientific community is to be cooperative as well as competitive. Writing research articles allows scientists to share knowledge. Scientists provide enough information so that other scientists can repeat the experiments or pertinent observations they describe. The journal *Science* along with several others presents its research articles in narrative form, and many of the details of the scientific method are understood rather than stated. However, adherence to the modern scientific method is expected, and the scientific community understands that it is as important to expose mistakes as it is to praise new knowledge.

MATERIALS

Per student:

- a typical research article in biology
- a research article from *Science*

PROCEDURE

1. Check the design of a typical research article in biology and list the titles of its various sections.

   (a) Example: Abstract (summary of paper)

   (b) _____

   (c) _____

   (d) _____

   (e) _____

   (f) _____

2. Read the article and then fill in the blanks in the following statements or answer the questions.

   (a) Any changes in the dependent variable are described in the _____ section.

   (b) Which section contains the details necessary to repeat this experiment or observation?
   _____

   (c) Which section contains the questions being asked, the predictions, or the hypotheses?
   _____

   (d) The _____ section contains the conclusions.

3. Look at an article from the journal *Science*. What steps of the scientific method are included in the narrative?
   _____
   _____

You no doubt know through various media sources of the myriad chemicals that pervade modern life. These substances can be beneficial or toxic to plants or animals. It's obvious that drinking gasoline is bad for you, but this is an extreme example. The broad questions are: what effects do new chemicals have on living organisms, and at what exposure level do they exert these effects? To answer these questions, we perform experiments known as bioassays. Bioassays are used by the pharmaceutical industry to test new drugs. Agricultural firms determine the effectiveness of new fertilizers and herbicides with bioassays. Some industries measure the effects their waste discharges have on aquatic organisms with bioassays. A **bioassay** establishes the quantity of a substance that results in a defined *effective dose* (ED)—that is, the dose that produces a particular effect. In the case of toxic substances, the standard measurement of toxicity is known as an LD_{50} (the lethal dose causing the death of 50% of the organisms exposed to the substance).

In this experiment you will test the hypothesis that many everyday substances affect germination of the seeds of Wisconsin Fast Plants™, members of the mustard family. Predictions are that if this is so, then the timing of seed germination will change, and if germination occurs, the extent of germination as measured by the length of roots and stems will change. If possible, you will use your results to estimate ED or LD_{50} for the substance tested.

MATERIALS

Per student:

- fine-pointed, water-resistant marker ("Sharpie" or similar)
- 35-mm film canister with lid prepunched with four holes or microcentrifuge tube holder (rack)
- four microcentrifuge tubes with caps
- four paper-towel wicks
- disposable pipet and bulb
- forceps
- eight RCBR seeds (Wisconsin Fast Plants™)
- 15-cm plastic ruler

Per student group (6):

- solution of test substance (paint thinner, household cleaner, perfume, coffee, vinegar, and so on)
- distilled water (dH₂O) in dropping bottle

CAUTION
Test substances may be harmful to you. Wear lab gloves and, even if you wear contact lenses, wear safety goggles. Eyeglasses should suffice when goggles do not fit over them.

PROCEDURE

1. Put on a pair of lab gloves and protective goggles.
2. Using a fine-pointed, water-resistant marker, label the caps and sides of the four microcentrifuge tubes 1.0, 0.1, 0.01, and C (for "Control").
3. Place 10 drops of the test solution into the tube labeled 1.0, which now contains 100% of the test solution. A test solution can be anything you and your instructor decide to test. Write the name of your test substance.
4. Make *serial dilutions* of the solution, producing concentrations of 10% and 1% of the test substance (Figure 1-5). Start by adding nine drops of dH₂O to the tubes labeled 0.1 and 0.01.
5. Remove a small quantity of the solution from tube 1.0 with the disposable pipet and place one drop into tube 0.1. Mix the contents thoroughly by flicking the sides of the bottom of the tube with the index finger of your dominant hand while holding the sides of the top of the tube between the index finger and thumb of your other hand. Return any solution remaining in the pipet to tube 1.0. Tube 0.1 now contains a concentration 10% of that in tube 1.0.
6. Now, from tube 0.1, remove a small quantity of the solution with the disposable pipet and place one drop in tube 0.01. Mix the contents thoroughly and return any solution left in the pipet to tube 0.1. Tube 0.01 now contains a 1% concentration of the original.
7. From the dropping bottle, place 10 drops of dH₂O into tube C. This is the control for the experiment; it contains none of the test solution.
8. Insert a paper-towel wick into each microcentrifuge tube (Figure 1-6). With your forceps, place two RCBR seeds near the top of each wick. Close the caps of the tubes. Do not allow any of the wick to protrude outside the cap.
9. Carefully insert the microcentrifuge assay tubes through the holes in the film canister lid (Figure 1-7). Set the experiment aside in the location indicated by your instructor.
10. At this point, write a prediction for your experiment in Table 1-3.

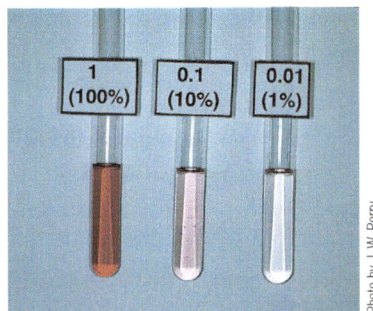

Figure 1-5 Serial dilution of test substance.

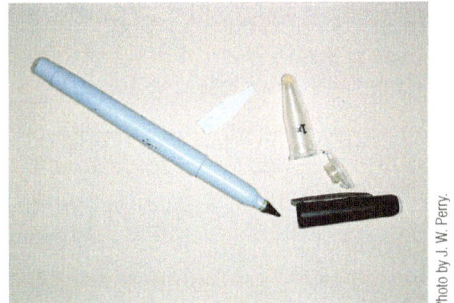

Figure 1-6 One of the four assay tubes.

Figure 1-7 Experimental apparatus.

TABLE 1-3	Effect of _____ on Germination of RCBR™ Seeds							

Prediction: _____

		My Experiment				Class Averages	
		Roots (mm)		Shoots (mm)			
Tube	Germination	S1	S2	S1	S2	Roots	Shoots
C (dH₂O)							
1.0							
0.1							
0.01							

© Cengage Learning 2013

11. State your experimental variables.

    Independent variable:

    _____

    _____

    Dependent variable:

    _____

    _____

    Controlled variables:

    _____

    _____

12. After 24 hours or more, examine your experiment to see how many and which seeds germinated. Record your results in Table 1-3. Record seeds as "germinated" (G) or "did not germinate" (DNG), and for germinated seeds indicate the extent of germination by measuring and recording the length in millimeters (mm) of the roots and shoots of seed 1 (S1) and seed 2 (S2) for each tube.

13. You have studied the effect of your substance on only a small number of seeds. To make your analysis more reliable, you should analyze a larger number of seeds. In an effort to increase reliability, pool your data with those students who tested the same substance. Calculate the averages and record them in Table 1-3.

14. If germination occurred in more than one group, graph the results in Figure 1-8, plotting the growth of roots and shoots at each concentration.

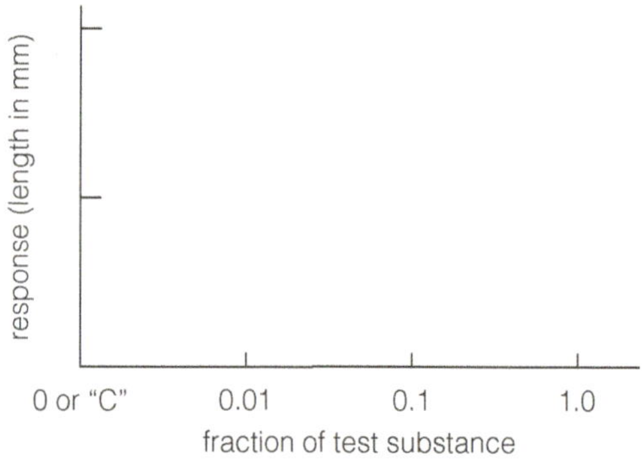

15. If germination occurred in the lower concentrations but not the higher ones, what effective dose prevents germination? Alternatively, could you have defined any other effective doses?

    _____

    _____

Figure 1-8 Effect of _____ on germination of RCBR seeds (O = roots and ● = shoots).

16. Considering your germination results, estimate or make a statement about the LD_{50} for your test substance. If negative effects on germination were observed, describe how you could modify the procedure to better determine the LD_{50}.

    _____

    _____

17. Make a conclusion regarding the effect of the test substance on germination of RCBR seeds.

_____

_____

18. Does your conclusion allow you to accept or reject your prediction? Is the hypothesis supported or falsified? Explain your answer.

_____

_____

_____ 1. The natural philosophy of Aristotle and his colleagues was
   (a) mechanistic.
   (b) vitalistic.
   (c) a belief in absolute truth.
   (d) a and c

_____ 2. A person who believes that the universe is at least partially controlled by supernatural powers can best be described as a(n)
   (a) teleologist.
   (b) vitalist.
   (c) empiricist.
   (d) mechanist.

_____ 3. The first step of the scientific method is to
   (a) ask a question.
   (b) construct a hypothesis.
   (c) observe carefully.
   (d) formulate a prediction.

_____ 4. Which series of letters lists the first four steps of the scientific method (see question 3) in the correct order?
   (a) a, b, c, d
   (b) a, b, d, c
   (c) c, a, b, d
   (d) d, c, a, b

_____ 5. In an experiment the subjects or items being investigated are divided into the experimental group and
   (a) the nonexperimental group.
   (b) the control group.
   (c) the statistics group.
   (d) none of these choices

_____ 6. The variables that investigators try to keep the same for both the experimental and the control groups are
   (a) independent.
   (b) controlled.
   (c) dependent.
   (d) a and c

_____ 7. Variables that are always different between the experimental and the control groups are
   (a) independent.
   (b) controlled.
   (c) dependent.
   (d) a and c

_____ 8. The results of an experiment
   (a) don't have to be repeatable.
   (b) should be repeatable by the investigator.
   (c) should be repeatable by other investigators.
   (d) must be both b and c.

_____ 9. The detailed report of an experiment is usually published in a
   (a) newspaper.
   (b) book.
   (c) scientific journal.
   (d) magazine.

_____ 10. Bioassays can be used to
   (a) test new drugs.
   (b) determine the effectiveness of new fertilizers and herbicides.
   (c) measure the effects their waste discharges have on aquatic organisms.
   (d) do all of these choices.

EXERCISE **1**

The Scientific Method

Post-Lab Questions

Introduction

1. How does the modern scientific method differ from the natural philosophy of the ancient Greeks?

1.1 Modern Scientific Method

2. List the six steps of one full cycle of the scientific method.

   (a) _____

   (b) _____

   (c) _____

   (d) _____

   (e) _____

   (f) _____

3. What is tested by an experiment?

4. Within the framework of an experiment, describe the

   (a) independent variable:

   (b) dependent variable:

   (c) controlled variables:

5. Is the statement, "In most biology experiments, the relationship between the independent and the dependent variable can best be described as cause and effect," true or false? Explain your answer.

6. Is a scientific principle taken as absolutely true? Explain your answer.

1.2 Research Article

7. What is the function of research articles in scientific journals?

1.3 Bioassay

8. Define the design and structure of a bioassay.

Food for Thought

9. Describe how you have applied or could apply the scientific method to an everyday problem.

10. Do you think the differences between religious and scientific knowledge make it difficult to debate points of perceived conflict between them? Explain your answer.

Measurement

OBJECTIVES

After completing this exercise, you will be able to

1. define *length, volume, meniscus, mass, density, temperature, thermometer;*

2. recognize graduated cylinders, beakers, Erlenmeyer flasks, different types of pipets, and a triple beam balance;

3. measure and estimate length, volume, and mass in metric units;

4. explain the concept of temperature;

5. measure and estimate temperature in degrees Celsius;

6. explain the advantages of the metric system of measurement.

Introduction

One requirement of the scientific method is that results be repeatable. As numerical results are more precise than purely written descriptions, scientific observations are usually made as measurements. Of course, sometimes a written description without numbers is the most appropriate way to describe a result.

| 2.1 | Metric System *(About 60 min.)* |

Metric system measurements are easy to convert from one unit to another (e.g., centimeters to meters to kilometers for length) and from one related unit to another (e.g., length to area to volume). The metric system is preferred by most of the people and countries in the world, especially by scientists. Governments have largely switched to it, at least in globally traded and shared manufacturing items (e.g., all cars made in the United States have metric parts and the U.S. Department of Defense adopted the metric system in 1957).

The metric reference units are the **meter** for length, the **liter** for volume, the **gram** for mass, and the **degree Celsius** for temperature. Regardless of the type of measurement, the same prefixes are used to designate the relationship of a unit to the reference unit. Table 2-1 lists the prefixes we will use in this and subsequent exercises.

TABLE 2-1	Prefixes for Metric System Units
Prefix of Unit (Symbol)	**Part of Reference Unit**
nano (n)	$1/1{,}000{,}000{,}000 = 0.000000001 = 10^{-9}$
micro (µ)	$1/1{,}000{,}000 = 0.000001 = 10^{-6}$
milli (m)	$1/1000 = 0.001 = 10^{-3}$
centi (c)	$1/100 = 0.01 = 10^{-2}$
kilo (k)	$1000 = 10^{3}$

© Cengage Learning 2013

As you can see, the metric system is a decimal system of measurement. Metric units are 10, 100, 1000, and sometimes 1,000,000 or more times larger or smaller than the reference unit. Thus, it's relatively easy to convert from one measurement to another either by multiplying or dividing by 10 or a multiple of 10:

$$
\text{nano} \left\{ \begin{array}{l} \text{meter} \\ \text{liter} \\ \text{gram} \end{array} \right.
\xleftarrow[\div 1000]{\times 1000 \quad \text{micro}}
\left\{ \begin{array}{l} \text{meter} \\ \text{liter} \\ \text{gram} \end{array} \right.
\xleftarrow[\div 1000]{\times 1000 \quad \text{milli}}
\left\{ \begin{array}{l} \text{meter} \\ \text{liter} \\ \text{gram} \end{array} \right.
\xleftarrow[\div 10]{\times 10 \quad \text{centi}}
\left\{ \begin{array}{l} \text{meter} \\ \text{liter} \\ \text{gram} \end{array} \right.
\xleftarrow[\div 100]{\times 100}
\begin{array}{l} \text{meter} \\ \text{liter} \\ \text{gram} \end{array}
$$

In this section, we examine the metric system and compare it to the American Standard system of measurement (feet, quarts, pounds, and so on).

MATERIALS

Per student pair:

- 30-cm ruler with metric and American (English) Standard units on opposite edges
- nonmercury thermometer(s) with Celsius (°C) and Fahrenheit (°F) scales (about –20°–110°C)
- 250-mL beaker made of heat-proof glass
- hot plate
- 250-mL Erlenmeyer flask
- three boiling chips
- three graduated cylinders: 10-mL, 25-mL, 100-mL
- thermometer holder

Per student group:

- a triple beam balance

Per lab room:

- source of distilled water (dH_2O)
- metric bathroom scale
- source of ice
- 1-quart jar or bottle marked with a fill line
- one-piece plastic dropping pipet (not graduated) or Pasteur pipet and bulb
- graduated pipet and safety bulb or filling device
- 1-pound brick of coffee
- ceramic coffee mug
- 1-gallon milk bottle
- metric tape measure
- 1-L measuring cup

PROCEDURE

A. Length (15 min.)

Length is the measurement of a real or imaginary line extending from one point to another. The standard unit is the meter, and the most commonly used related units of length are:

1000 millimeters (mm) = 1 meter (m)
100 centimeters (cm) = 1 m
1000 m = 1 kilometer (km)

For orientation purposes, the yolk of a chicken egg is about 3 cm in diameter. Since the differences between these metric units are based on 10 or multiples of 10, it's fairly easy to convert a measurement in one unit to another.

1. For example, if you want to convert 1.7 km to centimeters, your first step is to determine how many centimeters there are in 1 km. Remember, like units may be cancelled.

$$\frac{100 \text{ cm}}{1 \text{ m}} \times \frac{1000 \text{ m}}{1 \text{ km}} = \frac{100{,}000 \text{ cm}}{1 \text{ km}}$$

The second step is to multiply the number by this fraction.

$$\frac{1.7 \text{ km}}{1} \times \frac{100{,}000 \text{ cm}}{1 \text{ km}} = 170{,}000 \text{ cm}$$

The last calculation can also be done quickly by shifting the decimal point five places to the right.

$$1.700000 \text{ km} = 170{,}000.0 \text{ cm}$$

Alternatively, to convert 1.7 km to centimeters, we add exponents.

$$\frac{1.7 \text{ km}}{1} \times \frac{10^2 \text{ cm}}{\text{m}} \times \frac{10^3 \text{ m}}{\text{km}} = 1.7 \times 10^5 \text{ cm} = 170{,}000 \text{ cm}$$

Using the method most comfortable for you, calculate how many millimeters there are in 4.8 m.

_____ mm

Now let's convert 17 mm to meters.

step 1 $\dfrac{1 \text{ m}}{100 \text{ cm}} \times \dfrac{1 \text{ cm}}{10 \text{ mm}} = \dfrac{1 m}{1000 \text{ mm}}$

step 2 $\dfrac{17 \text{ mm}}{1} \times \dfrac{1 \text{ m}}{1000 \text{ mm}} = 0.017 \text{ m}$

Alternatively, you can shift the decimal point three places to the left,

0017.0 mm = 0.017 m

or add exponents:

$$\dfrac{1.7 \text{ mm}}{1} \times \dfrac{10^{-2} \text{ m}}{\text{cm}} \times \dfrac{10^{-1} \text{ cm}}{\text{mm}} = 1.7 \times 10^{-3} \text{ m} = 0.017 \text{ m}$$

Calculate how many kilometers there are in 16 cm.

_____ km

2. Precisely measure the length of this page in centimeters to the nearest 10th of a centimeter with the metric edge of a ruler. Note that nine lines divide the space between each centimeter into 10 millimeters.

The page is _____ cm long.

Calculate the length of this page in millimeters, meters, and kilometers.

_____ mm _____ m _____ km

Now repeat the above measurement using the American Standard edge of the ruler. Measure the length of this page in inches to the nearest eighth of an inch.

_____ in

Convert your answer to feet and yards.

_____ ft _____ yd

Explain why it is much easier to convert units of length in the metric system than in the American Standard system.

_____

_____

_____

B. Volume (20 min.)

Volume is the space an object occupies. The standard unit of volume is the liter (L), and the most commonly used subunit, the milliliter (mL). There are 1000 mL in 1 liter. A chicken egg has a volume of about 60 mL.

The volume of a box is the height multiplied by the width multiplied by the depth. The amount of water contained in a cube with sides 1 cm long is 1 cubic centimeter (cc), which for all practical purposes equals 1 mL (Figure 2-1).

1. How many milliliters are there in 1.7 L?

_____ mL

How many liters are there in 1.7 mL?

_____ L

2. Use Figure 2-2 to recognize **graduated cylinders, beakers, Erlenmeyer flasks,** and the different types of **pipets.** Some of these objects are

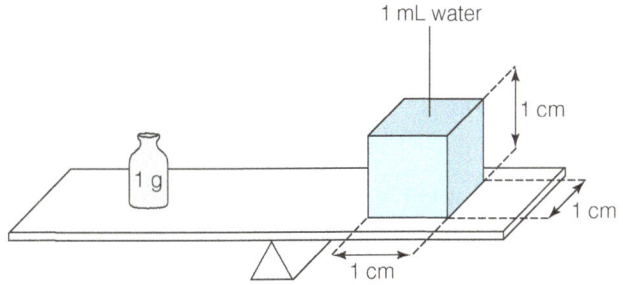

Figure 2-1 Relationship among the units of length, volume, and mass in the metric system.

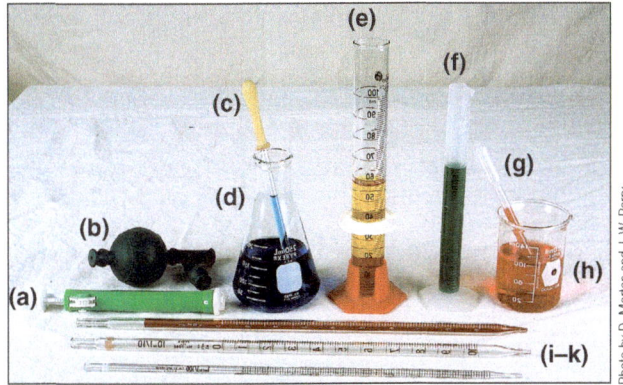

Figure 2-2 Apparatuses commonly used to measure volume: (**a**) pipet filling device, (**b**) pipet safety bulb, (**c**) Pasteur pipet, (**d**) Erlenmeyer flask, (**e**) glass graduated cylinder, (**f**) plastic graduated cylinder, (**g**) plastic dropping pipet, (**h**) beaker, (**i–k**) graduated pipets.

Photo by D. Morton and J. W. Perry.

made of glass; some are plastic. Some will be calibrated in milliliters and liters; others will not be.

3. Pour some water into a 100-mL graduated cylinder and observe the boundary between fluid and air, the **meniscus**. Surface tension makes the meniscus curved, not flat. The high surface tension of water is due to its cohesive and adhesive or "sticky" properties. Draw the meniscus in the plain cylinder outlined in Figure 2-3. The correct reading of the volume is at the *lowest* point of the meniscus.

4. Using the 100-mL graduated cylinder, pour water into a 1-quart jar or bottle. About how many milliliters of water are needed to fill the vessel up to the line?

_____ mL

5. Pipets are used to transfer small volumes from one vessel to another. Some pipets are not graduated (for example, Pasteur pipets and most one-piece plastic dropping pipets); others are graduated.
   (a) Fill a 250-mL Erlenmeyer flask with distilled water.
   (b) Use a plastic dropping pipet or Pasteur pipet with a bulb to withdraw some water.
   (c) Count the number of drops needed to fill a 10-mL graduated cylinder to the 1-mL mark. Record this number in Table 2-2.
   (d) Repeat steps b and c two more times and calculate the average for your results in Table 2-2.
   (e) Explain why the average of three trials is more accurate than if you only do the procedure once.

_____

_____

_____

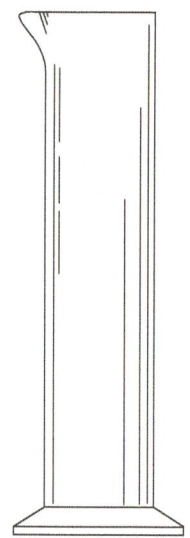

Figure 2-3
Draw a meniscus in this plain cylinder.

TABLE 2-2	Estimate of the Number of Drops in 1 mL
Trial	Drops/mL
1	
2	
3	
Total	
Average	

© Cengage Learning 2013

C. Mass (25 min.)

Mass is the quantity of matter in a given object. The standard unit is the **kilogram** (kg), and other commonly used units are the milligram (mg) and gram (g). There are 1,000,000 mg in 1 kg and 1000 g in 1 kg. A chicken egg has a mass of about 60 g. Note that the following discussion avoids the term *weight*. This is because weight depends on the gravitational field in which the matter is located. For example, you weigh less on the Moon but your mass remains the same as it is on Earth. Although it is technically incorrect, mass and weight are often used interchangeably.

1. How many milligrams are there in 1 g?

   _____ mg

   Convert 1.7 g to milligrams and kilograms.

   _____ mg _____ kg

2. A 1-cc cube, if filled with 1 mL of water, has a mass of 1 g (Figure 2-1). The mass of other materials depends on their **density** (water is defined as having a density of 1). The density of any substance is its mass divided by its volume.

   Approximately how many liters are present in 1 cubic meter (m^3) of water? As each of the sides of 1 m^3 are 100 cm in length, it's easy to calculate the number of cubic centimeters (that is, 100 cm × 100 cm × 100 cm = 1,000,000 cc). Now just change cubic centimeters to milliliters and convert 1,000,000 mL to liters.

   _____ L

What is its mass in kilograms?

_____ kg

In your "mind's eye," contemplate calculating how many pounds there are in a cubic yard of water. It's easier to convert between different units of the metric system than between those of the American Standard system.

3. Determine the mass of an unknown volume of water.* Mass may be measured with a **triple beam balance,** which gets its name from its three beams (Figure 2-4). A movable mass hangs from each beam.

(a) Slide all of the movable masses to zero. Note that the middle and back masses each click into the leftmost notch.

(b) Clear the pan of all objects and make sure it is clean.

(c) The balance marks should line up, indicating that the beam is level and that the pan is empty. If the balance marks don't line up, rotate the zero adjust knob until they do.

(d) Place a 250-mL beaker on the pan. The right side of the beam should rise. Slide the mass on the middle beam until it clicks into the notch at the 100-g mark. If the right end of the beam tilts down below the stationary balance mark, you have added too much mass. Move the mass back a notch. If the right end remains tilted up, additional mass is needed. Keep adding 100-g increments until the beam tilts down; then move the mass back one notch. Repeat this procedure on the back beam, adding 10 g at a time until the beam tilts down, and then backing up one notch. Next, slide the front movable mass until the balance marks line up.

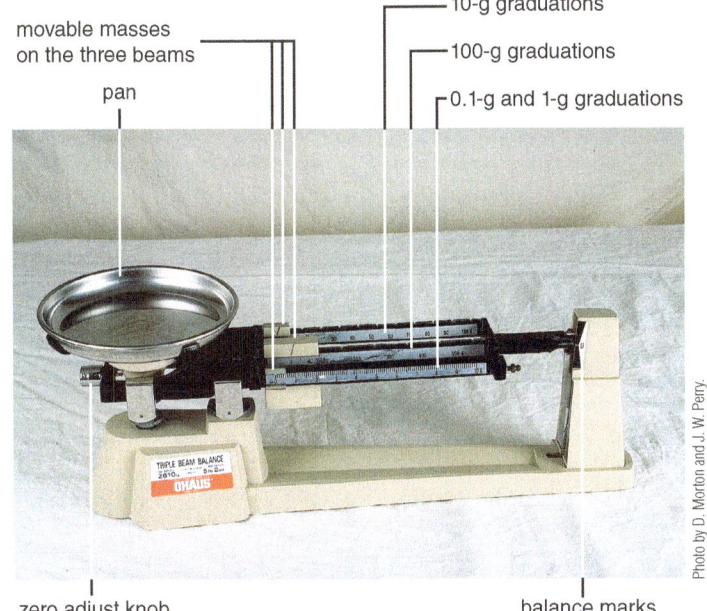

Figure 2-4 A triple beam balance.

(e) The sum of the masses indicated on the three beams gives the mass of the beaker. Nine unnumbered lines divide the space between the numbered gram markings on the front beam into 10 sections, each representing 0.1 g. Record the mass of the beaker to the nearest 10th of a gram in Table 2-3.

(f) Add an unknown amount of water and repeat the above procedure. Record the mass of the beaker and water in Table 2-3.

(g) Calculate the mass of the water alone by subtracting the mass of the beaker from that of the combined beaker and water. Record it in Table 2-3. Do not dispose of the water yet.

(h) Now measure the volume of the water in milliliters with a graduated cylinder. What is the volume? _____ mL

TABLE 2-3	Weighing an Unknown Quantity of Water with a Triple Beam Balance
Objects	**Masses (g)**
Beaker and water	
Beaker	
Water	

© Cengage Learning 2013

4. Using the triple beam balance, determine the mass of (that is, weigh) a pound of coffee in grams.
_____ g

---

*Modified from C. M. Wynn and G. A. Joppich, *Laboratory Experiments for Chemistry: A Basic Introduction*, 3rd ed. Wadsworth, 1984.

D. Estimating Length, Volume, and Mass (10 min.)

Now that you have experience using metric units, let's try estimating the measurements of some everyday items. Also, you may consult the metric/American Standard conversion table in Appendix 1 at the end of the lab manual.

1. Estimate the length of your index finger in centimeters. _____ cm
2. Estimate your lab partner's height in meters. _____ m
3. How many milliliters will it take to fill a ceramic coffee mug? _____ mL
4. How many liters will it take to fill a 1-gallon milk bottle? _____ L
5. Estimate the weight of some small personal item (for example, loose change) in grams. _____ g
6. Estimate your weight or your lab partner's weight in kilograms. _____ kg
7. Transfer your estimates to Table 2-4.
8. Then check your results using either a ruler, metric tape measure, 100-mL graduated cylinder, 1-L measuring cup, triple beam balance, or metric bathroom scale, recording your measurements in Table 2-4. Complete Table 2-4 by calculating the difference between each estimate and measurement.

TABLE 2-4	Differences Between Estimates and Measurements		
Number	**Estimate**	**Measurement**	**Estimate – Measurement**
1	cm	cm	cm
2	m	m	m
3	mL	mL	mL
4	L	L	L
5	g	g	g
6	kg	kg	kg

© Cengage Learning 2013

E. Temperature (About 20 min.)

The degree of hot or cold of an object is termed **temperature**. More specifically, it is the average kinetic energy of molecules. Heat always flows from high to low temperatures. This is why hot objects left at room temperature always cool to the surrounding or ambient temperature, while cold objects warm up. Consequently, to keep a heater hot and the inside of a refrigerator cold requires energy. **Thermometers** are instruments used to measure temperature.

1. Using a thermometer with both *Celsius* (°C) and *Fahrenheit* (°F) scales, measure room temperature and the temperature of cold and hot running tap water. Record these temperatures in Table 2-5.
2. Fill a 250-mL beaker with ice about three-fourths full and add cold tap water to just below the ice. Wait for 3 minutes, measure the temperature, and record it in Table 2-5. Remove the thermometer and discard the ice water into the sink.
3. Fill the beaker with warm tap water to about three-fourths full and add three boiling chips. Use a thermometer holder to clip the thermometer onto the rim of the beaker so that the bulb of the thermometer is halfway into the water. Boil the water in the beaker by placing it on a hot plate. After the water boils, record its temperature in Table 2-5. Turn off the hot plate and let the water and beaker cool to below 50°C before pouring the water into the sink.

> **CAUTION**
>
> Your instructor will give you specific instructions on how to set up the equipment in your lab for boiling water.

4. To convert Celsius degrees to Fahrenheit degrees, multiply by $\frac{9}{5}$ and add 32. Is 4°C the temperature of a hot or cool day? _____ What temperature is this in degrees Fahrenheit? _____ °F
5. To convert Fahrenheit degrees to Celsius degrees, subtract 32 and multiply by $\frac{5}{9}$. What is body temperature, 98.6°F, in degrees Celsius? _____ °C
6. In summary, the formulas for these temperature conversions are:
$$°F = \frac{9}{5} \, °C + 32$$
$$°C = \frac{5}{9} \, (°F - 32)$$

TABLE 2-5 Comparison of Celsius and Fahrenheit Temperatures

Location	°C	°F
Room		
Cold running tap water		
Hot running tap water		
Ice water		
Boiling water		

© Cengage Learning 2013

2.2 Micromeasurement *(About 30 min.)*

More and more procedures used in molecular biology and related areas require the measurement of very small masses and volumes. The following procedure introduces you to use of the electronic balance and micropipets or micropipetters. We'll do this as an experiment in part by answering the question: in the hands of students, how exact are instruments that deliver small volumes in doing what they say they do? Our hypothesis is that under these conditions these instruments perform within the manufacturer's stated specifications for the volumes they deliver.

MATERIALS

Per student group (table):

- standardized electronic balance capable of weighing 0.001 g and with at least a readability of 0.001 g, a repeatability (standard deviation) of 0.001 g, and a linearity of 0.002 g
- micropipets or micropipetters with matching tips (capable of delivering 10-μL, 50-μL, and 100-μL volumes)
- nonabsorbent weighing dish
- small test tube in a beaker or rack
- scientific calculator with instructions

Per lab room:

- source of distilled water (dH$_2$O)

PROCEDURE

1. Remember that in Section 2.1 we established a relationship between the volume and mass of water. If the mass of 1 μL of water is 1 g (Figure 2-1), how much does 1 μL weigh?

   _____ g or _____ mg

CAUTIONS ON THE USE OF *MICROPIPETTERS*

- Never adjust the micropipetter above or below the stated range or volume limit of measurement.
- Never use the micropipetter without a tip in place, as this will damage the internal piston mechanism.
- Never invert the micropipetter with a filled tip, as fluid could run up into its chamber.
- Never allow the plunger to snap back from the depressed position, as this could damage the piston. After drawing up or dispensing the fluid, release the plunger slowly.
- Never immerse the barrel of the micropipetter in fluid.

The electronic balances have been standardized prior to lab. If the micropipets or micropipetters in the hands of students perform within manufacturer's specifications, then we can predict what 100 μL, 50 μL, and 10 μL delivered from one of these instruments will weigh. In Table 2-6, write a prediction for this experiment.

2. Fill your test tube with distilled water.
3. Turn on the electronic balance and wait a moment (usually about 5 seconds) until it is stable.
4. Place a weighing dish on the pan.
5. Press the zero or tare key of the balance. Note that the digital display reads all zeros.
6. Use a 100-μL micropipet or a micropipetter adjusted to 100 μL and deliver this volume into the weighing dish. Wait a second for the balance to stabilize, then read and record the mass in grams or milligrams in

the correct column of Table 2-6. If you are using a toploading balance, the last digit may continue to change. In this case, use the middle number of the range of numbers displayed.

7. Tare or zero your balance and repeat step 6 for a total of 10 trials.
8. Tare or zero your balance and repeat steps 6 and 7 for the 50- and 10-μL volumes.
9. Use a scientific calculator to determine the average or mean and standard deviation for each column.
10. Plot your points on the graph (Figure 2-5a) and the means ± standard deviations on the bar chart (Figure 2-5b).

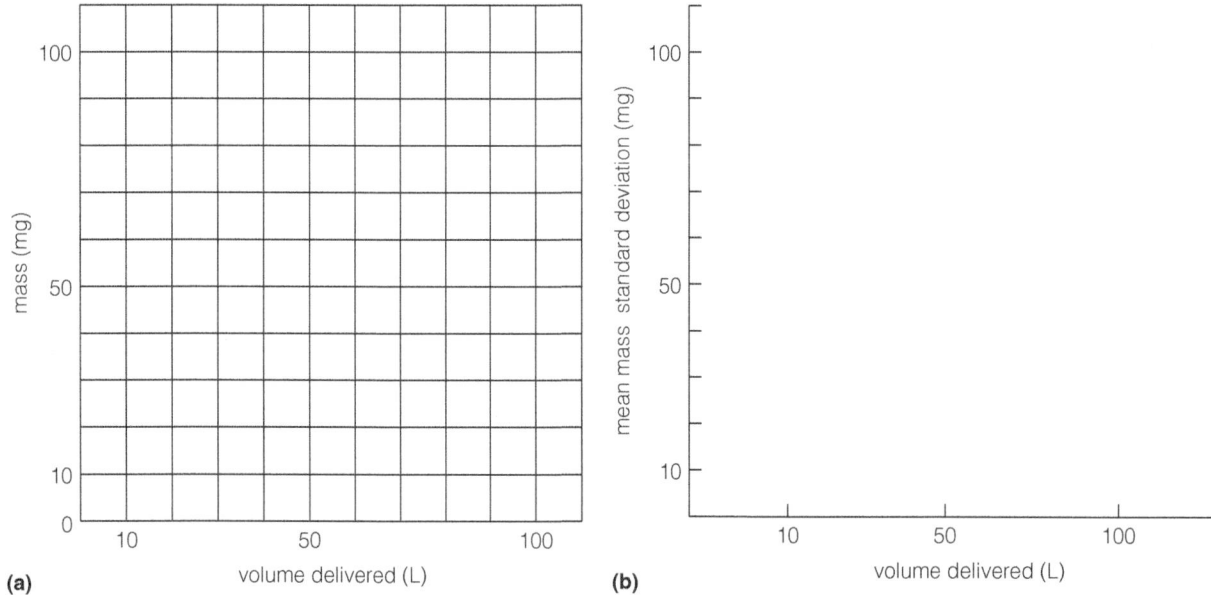

(a)

(b)

Figure 2-5 Micromeasurement results.

11. Write your conclusion in Table 2-6.

Trial	100 μL	50 μL	10 μL
1			
2			
3			
4			
5			
6			
7			
8			
9			
10			
Mean			
Standard deviation			

TABLE 2-6 Comparison of Volumes of Water with Their Masses

Conclusion:

© Cengage Learning 2013

_____ 1. The metric system is the measurement system of choice for
   (a) scientists.
   (b) most countries in the world.
   (c) most peoples of the world.
   (d) all of the above

_____ 2. A kilowatt, a unit of electrical power, is
   (a) 10 watts.
   (b) 100 watts.
   (c) 1000 watts.
   (d) 1,000,000 watts.

_____ 3. A millicurie, a unit of radioactivity, is
   (a) a tenth of a curie.
   (b) a hundredth of a curie.
   (c) a thousandth of a curie.
   (d) a millionth of a curie.

_____ 4. Length is the measurement of
   (a) a line, extending from one point to another.
   (b) the space an object occupies.
   (c) the quantity of matter present in an object.
   (d) the degree of hot or cold of an object.

_____ 5. Volume is the measurement of
   (a) a line, extending from one point to another.
   (b) the space an object occupies.
   (c) the quantity of matter present in an object.
   (d) the degree of hot or cold of an object.

_____ 6. Mass is the measurement of
   (a) a line, extending from one point to another.
   (b) the space an object occupies.
   (c) the quantity of matter present in an object.
   (d) the degree of hot or cold of an object.

_____ 7. If 1 cc of a substance has a mass of 1.5 g, its density is
   (a) 0.67.
   (b) 1.0.
   (c) 1.5.
   (d) 2.5.

_____ 8. If your mass is 70 kg on Earth, how much is your mass on the Moon?
   (a) less than 70 kg
   (b) more than 70 kg
   (c) 70 kg
   (d) none of the above

_____ 9. Above zero degrees, the actual number of degrees Celsius for any given temperature is _____ the degrees Fahrenheit.
   (a) higher than
   (b) lower than
   (c) the same as
   (d) a or b

_____ 10. A thermometer measures
   (a) the degree of hot or cold.
   (b) temperature.
   (c) a and b
   (d) none of the above

EXERCISE **2**

Measurement

Post-Lab Questions

Introduction

1. What is the importance of measurement to science?

2.1 Metric System

2. Convert 1.24 m to millimeters, centimeters, and kilometers.

_____ mm

_____ cm

_____ km

3. Observe the following carefully and read the volume.

_____ mL

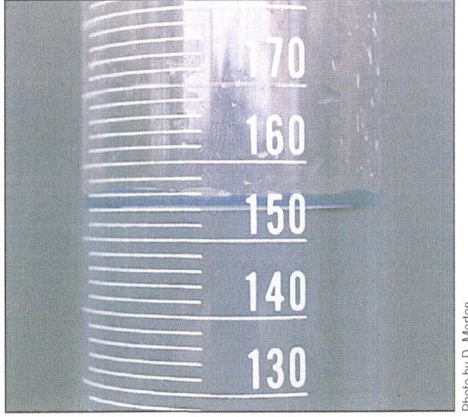

Photo by D. Morton.

4. Construct a conversion table for mass. Construct it so that if you wish to convert a measurement from one unit to another, you multiply it by the number at the intersection of the original unit and the new unit.

Original Unit	New Unit		
	mg	g	kg
mg	1		
g		1	
kg			1

5. How is it possible for objects of the same volume to have different masses?

6. Today, many packaged items have the volume or weight listed in both American Standard and metric units. Before your next lab period, find and list 10 such items.

Item	American Standard	Metric

7. Each °F is _____ (larger, smaller) than each °C.

2.2 Micromeasurement

8. Describe what it means to tare or zero an electronic balance. What purpose does this function serve?

Food for Thought

9. How are length, area, and volume related in terms of the three dimensions of space?

10. If you were to choose between the metric and American Standard systems of measurement for future generations, which one would you choose? Set aside your familiarity with the American Standard system and consider their ease of use and degree of standardization with the rest of the world.

Microscopy

Journi Norman (handwritten)

OBJECTIVES

After completing this exercise, you will be able to

1. define *magnification, resolving power, contrast, field of view, parfocal, parcentral, depth of field, working distance;*

2. describe how to care for a compound light microscope;

3. recognize and give the function of the parts of a compound light microscope;

4. accurately align a compound light microscope;

(handwritten: 10)

5. correctly use a compound light microscope;

6. make a wet mount;

7. correctly use a dissecting microscope;

8. describe the usefulness of the phase-contrast, transmission electron, and scanning electron microscopes;

9. use your skills to enjoy a fascinating world unavailable to the unaided eye.

Introduction

Light microscopes contain transparent glass lenses, which focus light rays emanating from a specimen to produce an image of a specimen on the retina, the light-sensitive layer of the eye. Table 3-1 presents the three most important properties of lenses and their images.

TABLE 3-1	Important Lens and Image Properties
Property	**Definition**
Magnification	The amount that the image of an object is enlarged (e.g., 100×)
Resolving power	The extent to which object detail in an image is preserved during the magnifying process
Contrast	The degree to which image details stand out against their background

© Cengage Learning 2013

3.1	**Compound Light Microscope** *(About 80 min.)*

In everyday life, the eye lens projects onto the retina a focused image of an object held no closer than about 10 cm. At this distance, details separated by 0.1 mm are visible. Most cells and related structures are smaller than this, and a light microscope is needed to see them. A microscope placed between the eye and a specimen (usually a section or thin object[s] mounted on a glass slide) acts to bring the specimen very close to the eye so you can focus on its details. Ultimately, the greater the proportion of the retina covered by the final image of the specimen, the greater its magnification. It does this by producing a series of magnified images.

Magnification without enough resolving power is referred to as empty, and with a light microscope, the maximum useful magnification is about 1000 times the diameter of the specimen (1000×). Above this value, additional details are missing. Furthermore, adequate contrast is needed to see the details preserved in an image. Dyes are usually added to sections of biological specimens to increase contrast.

Similar to automobiles, there are many models of compound light microscopes, and these instruments have numerous accessories that may or may not be present. Typical examples are shown in Figure 3-1, and one is diagrammed in Figure 3-2. If your microscope differs significantly from Figure 3-2, your instructor will give you an unlabeled diagram. If the instructor assigns you a specific microscope for your lab work, record its identification code in the second column of the first row in Table 3-2.

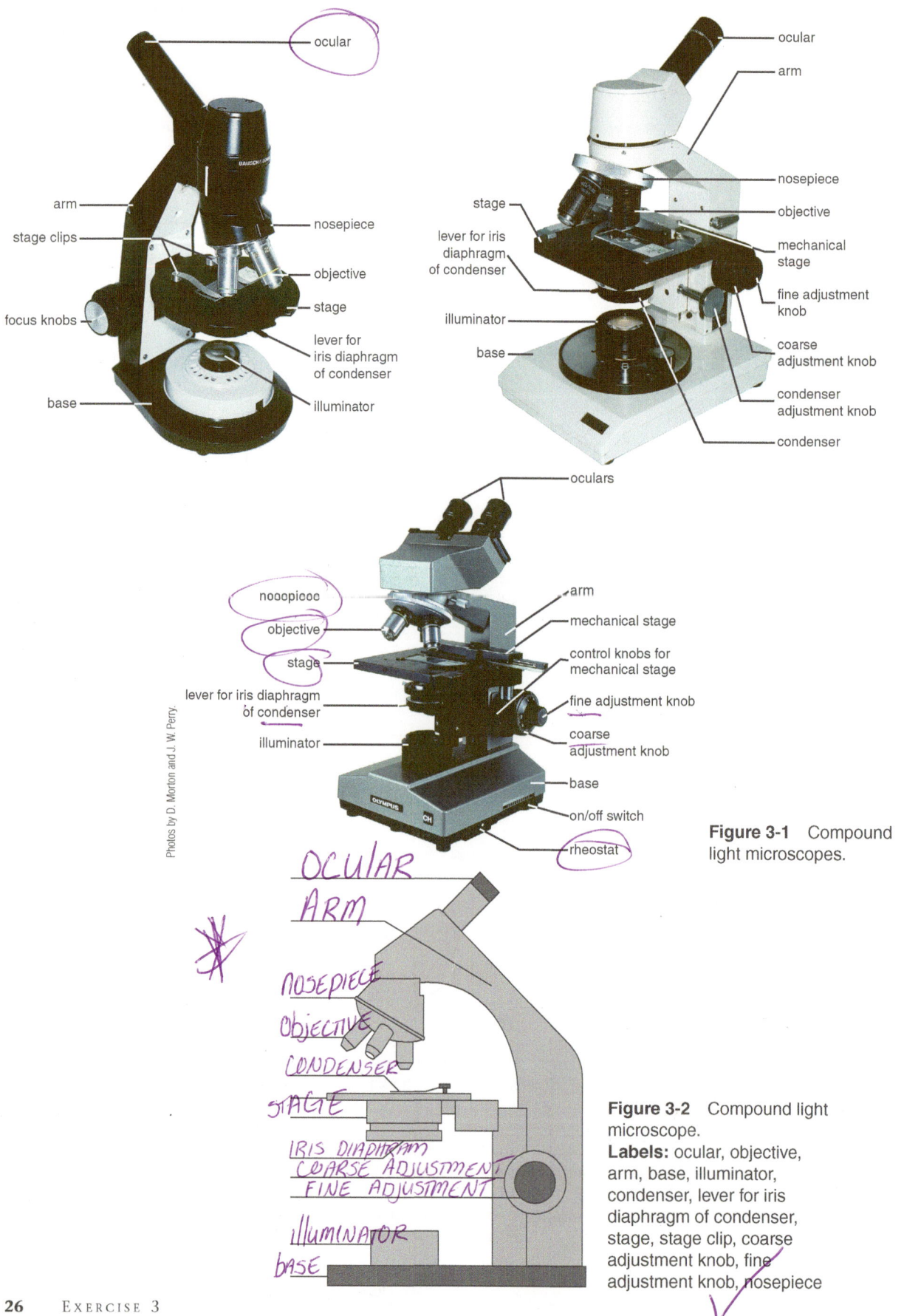

Photos by D. Morton and J. W. Perry.

Figure 3-1 Compound light microscopes.

Figure 3-2 Compound light microscope.
Labels: ocular, objective, arm, base, illuminator, condenser, lever for iris diaphragm of condenser, stage, stage clip, coarse adjustment knob, fine adjustment knob, nosepiece

TABLE 3-2	Characteristics of My Microscope	
Characteristic	Description	Function
Code	*numbers to id. scope*	identification of my microscope
Light source	*light under stage*	*to help see slides*
Condenser	*light hole on stage*	*control amount of light*
Stage	*flat platform*	*to place/hold slides*
Focusing knobs	*two knobs on right*	*to clearly see slides*
Objectives	*four lens*	*lens closest to slides*
Ocular(s)	*top two lens*	*lens magnify 10x*

© Cengage Learning 2013

MATERIALS

Per student:

- compound light microscope
- lens paper
- lint-free cloth (optional)
- unlabeled diagram of the compound light microscope model used in your course (optional)
- prepared slide with a whole mount of stained diatoms
- Wright-stained smear of mammalian blood
- prepared slide with the mounted letter "e"
- index card
- prepared slide with crossed colored threads coded for thread order

- prepared slide with a section of the mammalian kidney
- prepared slide with unstained fibers

Per student group (4):

- bottle of lens-cleaning solution (optional)
- dropper bottle of immersion oil (optional)

Per lab room:

- labeled chart of a compound light microscope

PROCEDURE

A. Care of a Compound Light Microscope

1. To carry a microscope to and from your lab bench, grasp the arm (Figure 3-1) with your dominant hand and support the **base** (Figure 3-1) with the other hand, always keeping the microscope upright. If the arm is wide, there should be a carrying indent or something similar to help you grasp it. *Do not try to carry anything else at the same time.* Label the arm and the base on Figure 3-2 or on the diagram given to you by your instructor.

> **CAUTION**
>
> **Never wipe a glass lens with anything other than lens paper.**

2. Remove the dust cover and clean the exposed parts of the optical system. Blow off any loose dust that may be on the ocular and then gently brush off any remaining dust with a piece of lens paper.

   If the lens of the ocular is still dirty, breathe on it and gently polish it with a rotary motion using a fresh piece of lens paper. If the ocular lens is still dirty, and *with your instructor's approval*, clean it with a piece of lens paper moistened with lens-cleaning solution.

3. Always remember that your microscope is a precision instrument. Never force any of its moving parts.

4. It is just as difficult to see clearly through a dirty slide as through a dirty microscope. Clean dirty slides with a lint-free cloth or with lens paper before using.

5. At the end of an exercise, make sure the last slide has been removed from the stage and *rotate the nosepiece so that the low-power objective is in the light path.* If your instrument focuses by moving the body tube, turn the coarse adjustment so that it is racked all the way down. If your microscope has an electric cord, neatly fold it up on itself and tie it with a plastic strap or rubber band. Otherwise, wind the cord around the base of the arm of the microscope.

6. Replace the dust cover before returning your microscope to the cabinet.

B. Parts of the Compound Light Microscope

Now that you know how to care for your microscope, remove the instrument assigned to you from the cabinet and place it on your lab bench. Use Figure 3-1 and the chart on the wall of your lab room to identify the various parts of your microscope. Read each step below and manipulate the parts *only where indicated*. Before you start, make sure the low-power objective is in the light path.

1. *Light source.* A compound microscope uses transmitted light to illuminate a transparent specimen usually mounted on a glass slide. Newer microscopes have a built-in **illuminator** (Figure 3-1). Locate the illuminator; the *off/on switch*; and perhaps also a rheostat, which is used to vary the intensity of the light. On some models, the switch and the rheostat are combined. Turn on the light source and look through the ocular. If the illuminator has a rheostat, adjust the intensity so the light is not too bright.

   (a) Label the illuminator on Figure 3-2 or on your instructor's diagram.

   (b) Describe the light source in Table 3-2, and then state its function.

2. *Condenser.* For maximum resolving power, a **condenser**—with a *condenser lens* and *iris diaphragm*—focuses the light source on the specimen so that each of its points is evenly illuminated. **The lever for the iris diaphragm** of the condenser is used to open and close the condenser. Instead of a lever, there may be a rotating ring.

   Establish whether there is a condenser adjustment knob (Figure 3-1) to set the height of the condenser. *Do not turn the knob; you will learn how to use it later.*

   There may be a *filter holder* under the condenser with a blue or frosted glass disk. Many microscope manufacturers believe that blue light is more pleasing to the eye because when used with an incandescent bulb, it produces a color balance similar to daylight conditions. Also, theoretically at least, blue light gives better resolving power because of its shorter wavelength. The frosted glass disk scatters light and can be useful in producing even illumination at low magnifications.

   (a) Label the condenser and lever for the iris diaphragm of the condenser on Figure 3-2 or on your instructor's diagram.

   (b) Describe the condenser and its related parts in Table 3-2 and then state its function.

3. *Stage.* Either a pair of *stage clips* or a *mechanical stage* holds a specimen mounted on a glass slide in place suspended over a central hole (Figure 3-1).

   (a) If your microscope has a mechanical stage, skip to part b. If your microscope has stage clips, place a prepared slide of stained diatoms under their free ends. Never remove the stage clips because they make it easier to move a slide in small increments. Skip b and do part c next.

   (b) Position a prepared slide of stained diatoms on the stage by releasing the tension on the spring-loaded movable arm of the mechanical stage (Figure 3-3a). There are two knobs to the right or left of the stage: one to move the specimen forward and backward and the other to move it

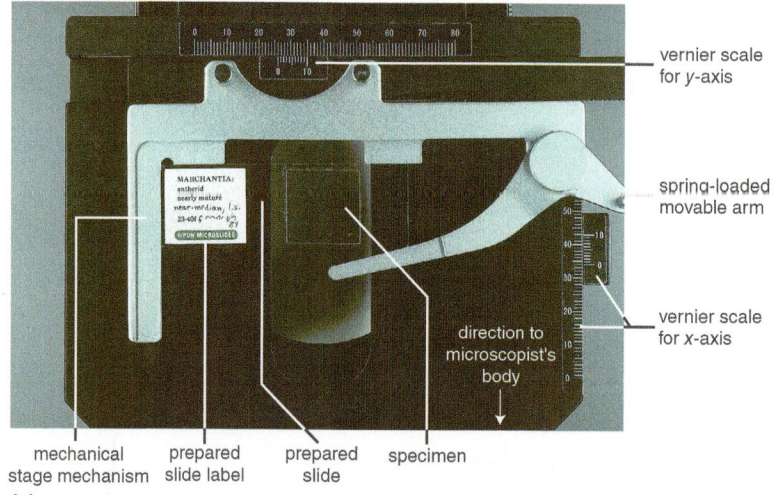

(a) — labels: vernier scale for *y*-axis; spring-loaded movable arm; vernier scale for *x*-axis; direction to microscopist's body; mechanical stage mechanism; prepared slide label; prepared slide; specimen

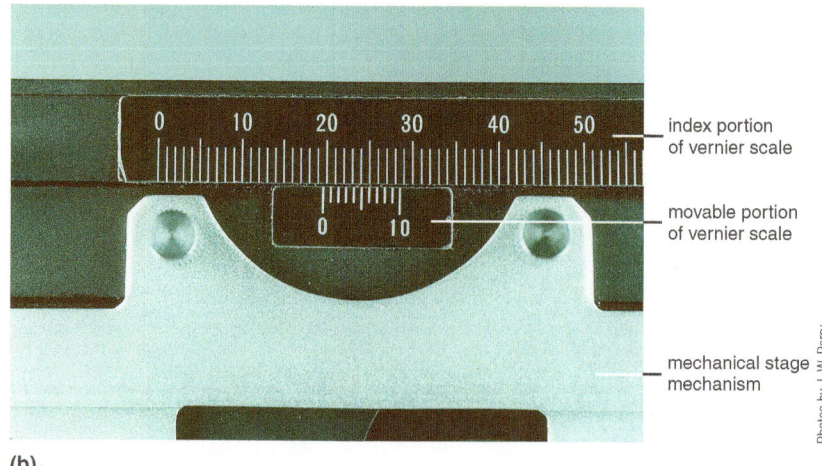

(b) — labels: index portion of vernier scale; movable portion of vernier scale; mechanical stage mechanism

Photos by J. W. Perry.

Figure 3-3 **(a)** Mechanical stage. **(b)** Vernier scale on mechanical stage of compound light microscope. The correct reading is 19.6 mm.

laterally (Figure 3-1). Label the stage and stage clips (or mechanical stage) on Figure 3-2 or on your instructor's diagram.

(c) On most mechanical stages, each direction has a vernier scale so you can easily locate interesting fields again and again. A vernier scale consists of two scales running side by side, a long one in millimeters and a short one that is 9 mm in length and divided into 10 equal subdivisions. To take a reading, note the whole number on the long scale coinciding with or just below the zero line of the short scale. If the whole number of the long scale and the zero of the short scale coincide, the first place after the decimal point is zero. Otherwise, the first place after the decimal point is the value of the line on the short scale that coincides (or nearly coincides) with one of the next nine lines after the whole number on the long scale. For example, the correct reading of the vernier scale in Figure 3-3b is 19.6 mm.

(d) Describe the stage and its related parts in Table 3-2 and then state its function.

4. *Focusing knobs* (Figure 3-1). The **coarse-focus adjustment knob** is for use with the lower-power objectives, and the **fine-focus adjustment knob** is for critical focusing, especially with the higher power objectives. On most modern microscopes, you move the stage of the instrument up and down to focus the specimen. Modern microscopes usually have a *preset focus lock*, which stops the stage at a particular height. After setting this lock, you can lower the stage with the coarse focus knob to facilitate changing of the specimen and then raise it to focusing height without fear of colliding the specimen against the objective. There may also be a *focus tension adjustment knob*, usually located inside of the left-hand coarse focus knob.

(a) Turn the coarse focus knob. Do you turn the knob toward you or away from you to bring the slide and objective closer together?

_____

(b) Label the coarse and fine-focus adjustment knobs on Figure 3-2 or on your instructor's diagram.

(c) Describe the focusing knobs in Table 3-2 and then state their function. As part of your description note if a preset focus lock or focus adjustment knob is present.

5. *Objectives.* The compound light microscope has at least two magnifying lenses, the objective and the ocular (Figure 3-1). The objective scans the specimen. Most microscopes have several objectives mounted on a revolving **nosepiece**. The magnifying power of each objective is labeled on its side. Usually included are these objectives: a 4× *low-power* or scanning, a 10× *medium-power* (Figure 3-4a), an about 40× *high-dry*, and perhaps an about 100× *oil-immersion objective* (Figure 3-4b). The other number often labeled on the side of nosepiece objectives is the **numerical aperture** (NA). The larger the NA is, the greater the resolving power and useful magnification.

MEASURE
of QUALITY
of lens.
(decimal)

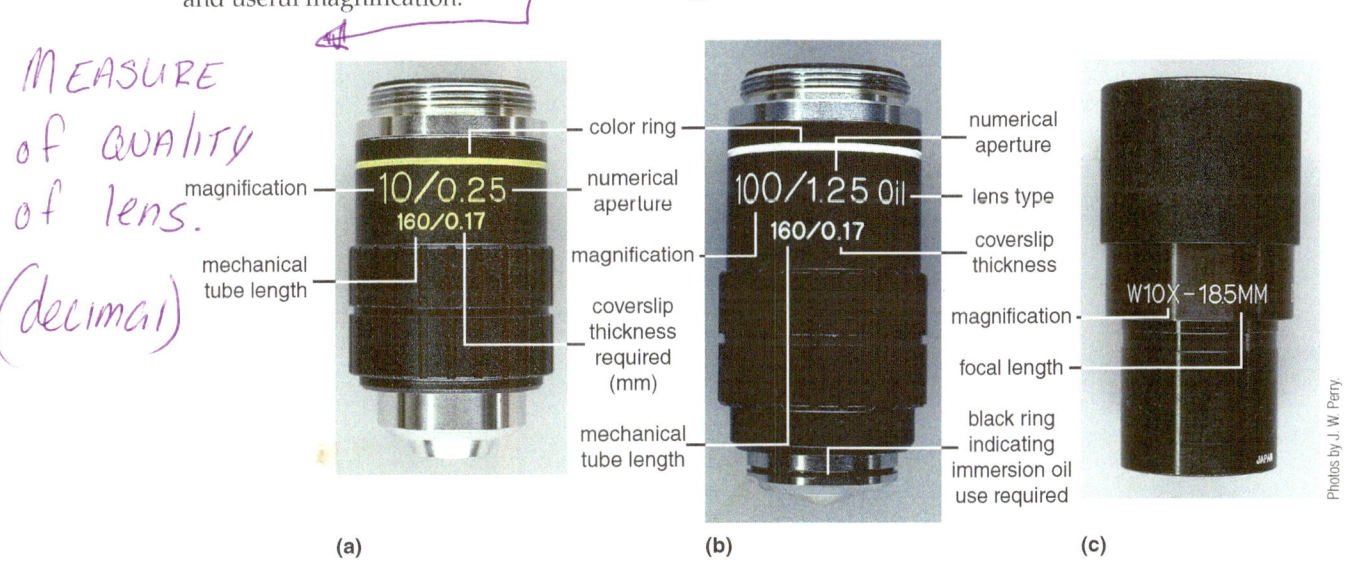

Figure 3-4 (**a**) 10× objective removed from microscope. (**b**) 100× oil-immersion objective removed from microscope. (**c**) Ocular removed from microscope.

In a properly aligned microscope, the objectives are **parfocal**. That is, when an objective has been focused, you can rotate to another one, and the image will remain in coarse focus, requiring only slight movement of the fine-focus knob. Objectives are also **parcentral**, meaning that the center of the field of view remains about the same for each objective. The **field of view** is the circle of light you see when looking into the microscope.

Often, objectives have different lengths; the lower-power objectives are shorter than the higher-power ones. That is, the **working distance** of objectives decreases with magnification. Working distance is the space between the objective lens and the slide. Therefore, the higher the power of the objective in use, the closer the objective is to the slide—and the more careful you must be.

(a) Record the magnifying power and NA of the objectives on your microscope in Table 3-3. If your instrument does not have a particular objective, indicate that it is not present (NP).

TABLE 3-3	Objectives Present on My Compound Light Microscope		
Objective	Objective Magnifying Power (ObMP×)	Total Magnifying Power (ObMP × OcMP = ___ ×)	Numerical Aperture (NA)
Low-power	4×		0.10
Medium-power	10×		0.25
High-dry	40×		0.65
Oil-immersion	100×		1.25

© Cengage Learning 2013

(b) Label the nosepiece and objective on Figure 3-2 or on your instructor's diagram.
(c) Describe the objectives in Table 3-2 and then state their function.
6. *Ocular.* The magnifying lens you look into is called an **ocular** (Figure 3-4c). Oculars are generally 10×.
(a) Because each objective has a different magnifying power, the total magnification is calculated by multiplying the magnifying power of the ocular by that of the objective in use. What is the ocular magnification power (OcMP) of the ocular(s) on your microscope? _____ ×
Calculate the total magnification for each of your microscope's ocular and objective combinations and then record them in Table 3-3.
(b) Label the ocular on Figure 3-2 or on your instructor's diagram.
(c) Describe the ocular in Table 3-2 and then state its function.
(d) Your microscope will have one or two oculars mounted *on a monocular or binocular head*, respectively. There may be a pointer mounted in an ocular so that you can easily show a specimen detail to your instructor or another student. For a monocular microscope, it's best to use your dominant eye to look down the ocular, keeping your other eye open. Is your microscope monocular or binocular? _____

(e) If your microscope is monocular, determine your dominant eye.
- Look at a small object on the far wall of your room with both eyes open.
- Form your thumb and index finger of one hand into a circle and place this circle in your line of sight at arm's length so it surrounds the object.
- Close your right eye. If the object shifts out of the circle to your left, your right eye is probably dominant. If the object remains in the circle, your left eye is probably dominant.   RIGHT
- This time, close your left eye and go through the process again. If the object shifts to the right, your left eye is dominant. If the object remains within the circle, your right eye is dominant. The more pronounced the shift, the greater the dominance. If there is no shift, neither eye is dominant.

C. Aligning a Compound Light Microscope with In-Base Illumination and a Condenser with an Iris Diaphragm

Aligning your microscope properly will not only help you see specimen detail clearly but will also protect your eyes from strain.

1. Rotate the nosepiece until the medium-power objective is in the light path. Open the iris diaphragm.
2. If it is not already there, place the prepared slide of stained diatoms on the stage; center and carefully focus on it. *Skip steps 3 and 4 if your microscope is monocular. Skip step 5 if your microscope does not have a control to adjust the height of the condenser.*
3. If your microscope is binocular, adjust the interpupillary distance. Hold a different ocular tube with each hand and while looking at the specimen, pull the tubes apart or push them together until you see one field of view. After making this adjustment, read and record the number off the scale.
My interpupillary distance is _____.
From now on, you can set the interpupillary distance at this number.

4. Now compensate for any difference in diopter between the lenses of each eye.
   (a) *If there is one diopter adjustment ring around the left ocular tube*, cover your left eye with an index card and focus your microscope using the fine-focus knob. Now uncover the left eye and cover your right one. Use the diopter adjustment ring to bring the specimen into focus.
   (b) *If both ocular tubes have a diopter adjustment ring*, set the left one to the same number as the interpupillary distance, cover your right eye with an index card, and focus on the specimen. Then uncover the right eye and cover your left one. Use the diopter adjustment ring on the right ocular tube to bring the specimen into focus.
5. Place a sharp point (pencil, dissecting needle, or some similar object) on top of the illuminator and bring the silhouette into sharp focus by adjusting the height of the condenser.
   (a) *If the ocular on your microscope is removable (and with the permission of your instructor)*, carefully slide it out and put the ocular open end down on a piece of lens paper in a safe place. Then, while looking down the ocular tube, adjust the iris diaphragm until the edge of the aperture lies just inside the margin of the back lens element of the objective (Figure 3-5). When done, replace the ocular.
   (b) *If the ocular cannot be removed*, close the condenser diaphragm and then open it until there is no further increase in brightness. Now close it again, stopping when you see the brightness begin to diminish.
   (c) Models that have rings that open and close the iris diaphragm of the condenser usually have numbers associated with them that match the magnification of the objective being used. Every time you switch objectives, you need to reset the ring to the correct position.
6. If your microscope has a rheostat, adjust the illumination to a level that lets you see specimen detail and that is comfortable for your eyes. To maintain the same illumination at higher magnifications, you will have to increase its intensity.
7. For best results, repeat steps 5 and 6 every time you use a different objective.

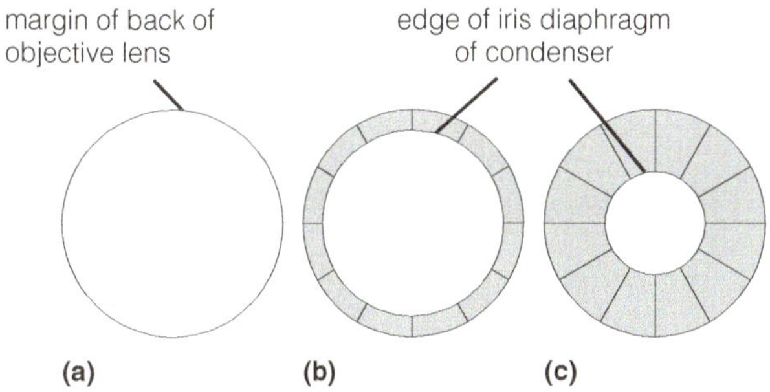

margin of back of objective lens

edge of iris diaphragm of condenser

(a)   (b)   (c)

Figure 3-5 Correct setting for condenser iris diaphragm. Drawing (**b**) is correct. In drawing (**a**), you cannot see the edge of the iris diaphragm. In (**c**), the diaphragm has been closed too much.

D. Using Different Magnifications

It is safest to observe a specimen on a slide first with low power and then, step by step, with higher power objectives. This way you avoid colliding the objective and slide together.

Because the magnification is in diameters, the area of the field of view decreases dramatically with increasing magnification (Figure 3-6). It follows that it is easier to use a lower power objective to locate a specific specimen detail. This is why the low-power objective is sometimes called the *scanning objective*. Also, if you lose a specimen detail at higher magnification, it is always easier to find it again if you switch to a lower power objective.

Now follow these steps to use each objective.

1. Rotate the low-power objective into the light path.
2. If it is not already there, place a prepared slide of stained diatoms on the stage, securing it with either the stage clips or the movable arm of the mechanical stage.
3. Look through the ocular. Bring the diatoms into focus using the *coarse focus knob*. Adjust the illumination as described in steps 5 and 6 of Section C. At this magnification, the diatoms appear small. Center a diatom by moving the slide.
4. Rotate the nosepiece so the medium-power objective is in the light path. Adjust the illumination. Focus the diatom.
5. Rotate the nosepiece so the high-dry objective is in the light path. Adjust the illumination. Focus the diatom using the *fine-focus adjustment knob*.

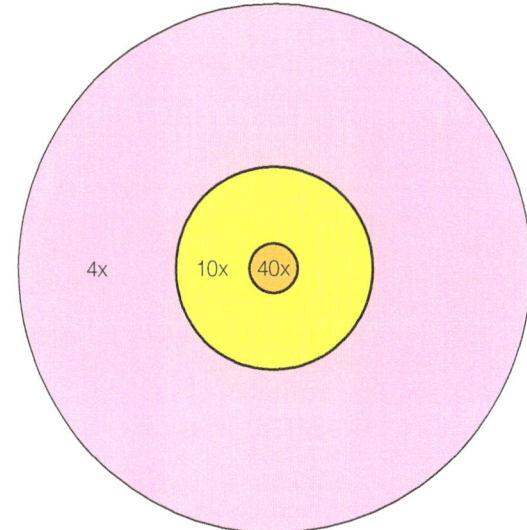

Figure 3-6 Illustration of the decreasing area of the field of view when a 4×, 10×, and 40× objective is used with a 10× ocular. The actual area of each circle has been enlarged 7.5×.

6. If your microscope has an oil-immersion objective and *with the permission of your instructor*, replace the prepared slide of diatoms with a smear of mammalian blood and repeat steps 1–5, otherwise skip steps 6–11 and just remove the prepared diatom slide from the stage. Center and focus on a prominent blue nucleus of a white blood cell. Most of the cells are red blood cells stained a light pink.
7. Rotate the nosepiece so that the light path is midway between the high-dry and oil-immersion objectives.
8. Place a small drop of immersion-oil on to the cover slip, using the circle of light above the specimen as a guide. Rotate the nosepiece so that the oil-immersion objective is in the oil.
9. Adjust the illumination and focus the white blood cell nucleus with the fine-focus knob.
10. When you are done examining the white blood cell nucleus, rotate the oil-immersion objective out of the light path, carefully wiping the oil from the oil-immersion objective with lens paper.
11. Remove the slide from the stage and wipe the cover slip with lens paper.

E. Orientation of the Image Compared with the Specimen

1. If you have not already done so, remove the prepared slide of diatoms and replace it with a prepared slide with the letter *e*. With the medium-power objective in the light path, position the slide with the specimen (the letter *e*) right side up on the stage. Center the *e* in the field of view and carefully bring it into focus.
2. In Figure 3-7, draw the image of the *e* as you see it through the ocular. Record the total magnification used in the line at the end of the legend.
   Is the image right side up or upside down compared with the specimen? _____RIGHT SIDE UP_____

   Compared with the specimen, is the image backward as well as upside down? ___NO___
   (yes or no)
   In summary, the image is *inverted* with respect to the specimen.
3. Move the specimen to the right while watching it through the microscope. In which direction does the image move? ____LEFT____

4. Move the specimen away from you. In which direction does the image move? ____UP, CLOSER____

5. Remove the slide and put it away.

Figure 3-7 Drawing of the letter *e* as seen through the ocular (_____ ×).

F. Depth of Field

The **depth of field** is the distance through which you can move the specimen and still have it remain in focus. Remember, the working distance—the space between the objective lens and the coverslip—decreases with increasing magnifying power. Therefore, the higher the power of the objective in use, the closer the objective is to the slide—and the more careful you must be.

1. Obtain a prepared slide of three crossed colored threads. *This exercise requires care because you are probably not yet adept at focusing on a specimen.* When you have the threads in focus (using first the low-power objective and then the medium-power objective), you need only use the fine-focus knob to focus with the high-dry objective. After switching to the high-dry objective, try rotating the fine-focus knob ½ turn away from you and then a full turn toward you. If you have not found the plane of focus, next try 1½ turns away from you and 2 full turns toward you and so on. If you work deliberately, you will find the plane of focus and will not crack the coverslip.
   (a) How many threads are in focus using the
       low-power objective? __0__
       medium-power objective? __1__
       high-dry objective? __3__
   (b) With which objective is it easiest to focus a specimen? __4 x__
   (c) At which magnification is it most difficult to focus a specimen? __40x__
2. Specimens have depth. Continue using the prepared slide of three crossed colored threads.
   (a) Use the high-dry objective to determine the order of the three threads mounted on the slide and record the results in Table 3-4. (Each slide label has a code on it. When you believe you have discovered the correct order, check with your instructor to find out if you are correct.)
   (b) Focusing carefully with the fine-focus knob, move from the bottom to the upper thread.
       Did you move the knob away from or toward you? __AWAY__

© Cengage Learning 2013

TABLE 3-4	Order of Threads
Location	**Color**
Closest to slide	
Middle	
Closest to coverslip	

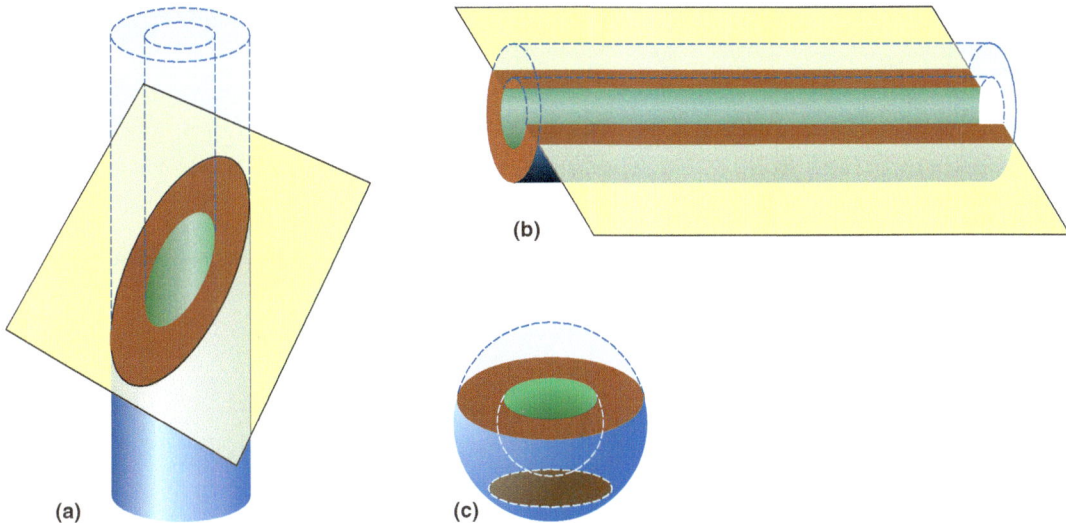

(b)

(a)   **(c)**

Figure 3-8 Difficulties encountered in interpreting the three-dimensional shape of objects from sections. Compare the transverse (cross) section of the cylinder with the distorted oblique section (**a**) and the longitudinal section (**b**). A similar shape to the transverse section of a cylinder results when a hollow ball is sectioned (**c**). A section through the wall of the hollow ball results in a solid shape.

3. Remove and put away the slide.
4. Viewing sections of three-dimensional structures makes interpretation of the original shape quite difficult.
   (a) Examine a slide of the cortex of the mammalian kidney with your compound microscope (Figure 3-8). Hypothesize as to the three-dimensional shape of the structures labeled *renal corpuscles* and *nephron tubules*.
   The shape of renal corpuscles is _____.
   The shape of nephron tubules is _____.
   (b) Check the textbook to see if your hypotheses are acceptable.

G. Using the Iris Diaphragm to Improve Contrast

1. Place a specimen of unstained fibers on the stage. Locate and focus on these fibers using the medium-power objective. Make sure the condenser and iris diaphragm are correctly set.
2. Close the iris diaphragm.
   Does this procedure increase or decrease contrast? _DECREASE_
   Although this procedure is useful when viewing specimens with low contrast, it should be used only as a last resort because resolving power is also decreased.
3. Remove and put away the slide.

H. Units of Measurement

The basic metric unit of length at the light-microscopic level is the micrometer (μm).

$1000 \ \mu m = 1 \ mm$
$1 \ \mu m = 0.001 \ mm$

How many nanometers are there in 1 mm? _____ nm
How many millimeters are there in 1 nm? _____ mm

In the mid-seventeenth century, Robert Hooke used a microscope to discover tiny, empty compartments in thin shavings of cork. He named them cells. Repeating this historic observation is a good way to learn how to prepare a wet mount.

MATERIALS

Per student:

- compound microscope, lens paper, a bottle of lens-cleaning solution (optional), a lint-free cloth (optional)
- cork
- razor blade

- glass microscope slide
- glass coverslip
- dissecting needle

Per student group (4):

- dropper bottle of distilled water (dH$_2$O)

PROCEDURE

1. Carefully use a razor blade to cut a number of *very thin shavings* from a cork stopper. Place them on a glass microscope slide.
2. Gently add a drop of distilled water.
3. Place one end of a glass coverslip to the right or left of the specimen so that the rest of the slip is held at a 45-degree angle over the specimen (Figure 3-9a).

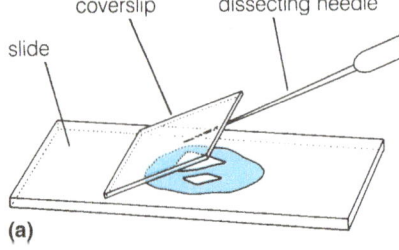
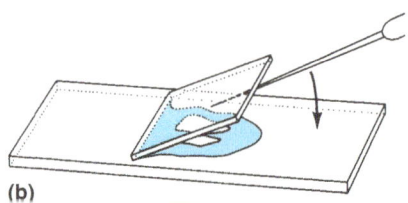

coverslip   dissecting needle

slide

(a)   (b)

Figure 3-9 How to make a wet mount.

4. Slowly lower the coverslip with a dissecting needle so as not to trap air bubbles (Figure 3-9b).
5. Observe the wet mount, first at low magnification and then with higher power. Air may be trapped either in the cork or as free bubbles (Figure 3-10). Trapped air will appear dark and refractive around its edges. This effect is caused by sharply bending rays of light. Draw what you see in Figure 3-11. Note the total magnification used to make the drawing.
6. Clean and replace the slide and coverslip as indicated by your instructor.

Photo by J. W. Parry.

Figure 3-10 Free air bubble (250×).

Figure 3-11 Drawing of the microscopic structure in a cork shaving (_____ ×).

Examining the microscopic world is both challenging and fun. Most of the macroscopic world has been explored, but the microscopic world is barely touched. Yet the microbes in it are essential to our very existence. So be an explorer and see what you can discover!

MATERIALS

Per student:

- compound microscope, lens paper, a bottle of lens-cleaning solution (optional), a lint-free cloth (optional)
- glass microscope slide
- glass coverslip

Per student group (4):

- pond water or some other mixed culture in a dropper bottle
- dropper bottle of Protoslo®

Per lab room:

- reference books for the identification of microorganisms

PROCEDURE

1. Obtain a drop of pond water or other mixed culture from the bottom of the bottle.
2. Add a drop of Protoslo®. This methyl cellulose solution slows down any swimming microorganisms.
3. Make a wet mount (Figure 3-9).
4. Observe the wet mount with your compound microscope. Start at the upper left corner of the coverslip and scan the wet mount with the low-power objective. When you find something interesting, focus on it and switch to the medium-power objective and then, if necessary, the high-dry objective.
5. Draw what you find on Figure 3-12 and note the total magnification.
6. Attempt to identify what you found using the resource books provided by your instructor. If successful, write its name under your drawing.
7. Clean and replace the slide and coverslip as indicated by your instructor.
8. Put away your compound microscope as described on page 27.

Figure 3-12 Drawing of microorganisms (_____ ×).

| 3.4 | **Dissecting Microscope *(About 10 min.)*** |

Dissecting microscopes (Figure 3-13) have a large working distance between the specimen and the objective lens. They are especially useful in viewing larger specimens (including thicker slide-mounted specimens) and in manipulating the specimen (e.g., when dissection of a small structure or organism is required).

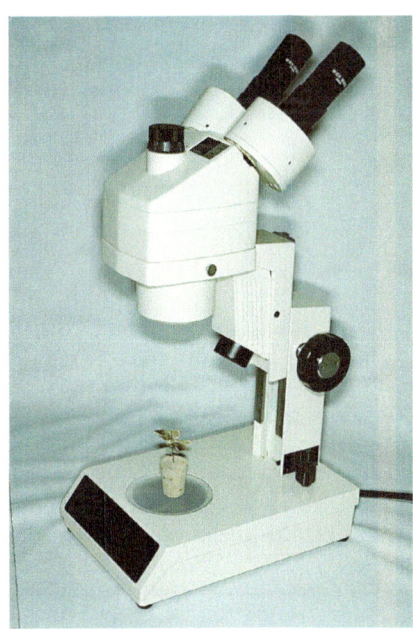

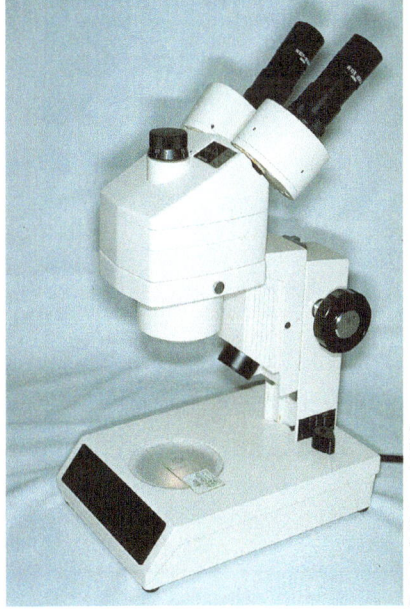

Figure 3-13 Dissecting microscope (**a**) reflected light, (**b**) transmitted light.

(a)

(b)

Photos by D. Morton and J. W. Perry

The large working distance also allows for illumination of the specimen from above (reflected light) as well as from below (transmitted light). Light reflected from the specimen shows surface features better than transmitted light.

MATERIALS

Per student group:

- dissecting microscope
- specimens appropriate for viewing with the dissecting microscope (e.g., a prepared slide with a whole mount of a small organism, bread mold, an insect mounted on a pin stuck in a cork, a small flower)

PROCEDURE

1. Under a dissecting microscope, view one or more of the specimens provided by your instructor. What is the magnification range of this microscope?

   _____ × to _____ ×

2. Is the image of the specimen inverted as in the compound light microscope? (yes or no) _____
3. Describe the type of illumination used by your dissecting microscope. Is there a choice?

   _____

   _____

3.5   Other Microscopes *(About 10 min.)*

In future exercises, you will examine pictures taken with other types of microscopes. Some will be of living cells taken with a phase-contrast microscope (Figure 3-14a) or similar instrument, including those using the Nomarski process (Figure 3-14b). Others will be of very thin-sectioned, heavy metal–stained specimens taken with a transmission electron microscope or TEM (Figure 3-14c). Still others will be of precious metal–coated surfaces produced by signals from the scanning electron microscope or SEM (Figure 3-14d). Table 3-5 summarizes the technology and use of these microscopes.

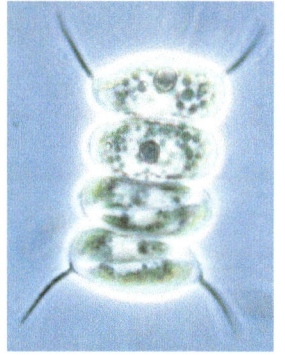 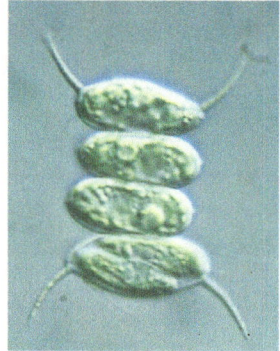 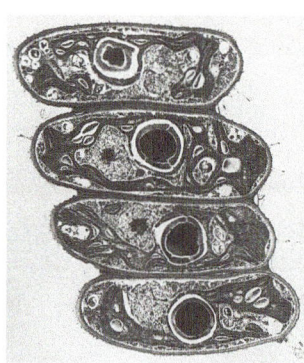 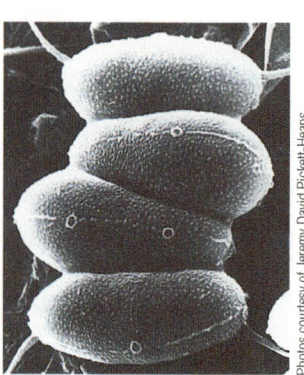

(a) phase contrast   **(b)** Nomarski process   **(c)** transmission electron   **(d)** scanning electron

Photos courtesy of Jeremy David Pickett-Heaps.

Figure 3-14   How different types of microscopes reveal detail in cells of the green alga *Scenedesmus*.

TABLE 3-5	Other Microscopes	
Microscope	**Technology**	**Use**
Phase-contrast and Nomarski process	Converts phase differences in light to differences in contrast	Observation of low-contrast specimens (often living)
Transmission electron	Increases resolving power by using electrons in a vacuum and magnetic lenses instead of light and glass lenses, respectively	Preservation of greater specimen detail allows for magnifications up to 1,000,000× or more (usually dead materials)
Scanning electron	Forms a TV-like picture from a secondary electron signal, which is emitted from surface points excited by a thin beam of electrons drawn across	Investigation of the fine structure of surfaces (usually dead specimens); the surface in a raster pattern

© Cengage Learning 2013

MATERIALS

Per student group:

- photographs of TEM micrographs (negatives) or digital images
- photographs of SEM negatives or digital images

PROCEDURE

1. Examine some photographs of TEM micrographs (negatives) or digital images. The darker areas are more electron dense in the specimen than the lighter areas.
2. Now look at some photographs of SEM negatives or digital images. The lighter areas correspond to the emission of greater numbers of secondary electrons from that part of the specimen; the darker areas emit fewer secondary electrons.
3. What type of microscope (compound light, dissecting, phase-contrast, TEM, or SEM) would you use to examine the specimens listed in Table 3-6?

TABLE 3-6 Microscope Use	
Specimen	Microscope
Living surface of the finger	
Dye-stained slide of a section of the finger	
Gold-coated bacteria on a single cell of the finger	
Unstained section of a biopsy from the finger	
Heavy metal–stained, very thin section of the finger	

© Cengage Learning 2013

FORMULA:

FINDING WIDTH of light

$$4.5mm \times \frac{\text{TOTAL MAG low power}}{\text{TOTAL MAG } 40\times} = \frac{4.5mm \times 40\times}{400\times}$$

$$= 0.45mm$$

_____ 1. Magnification
   (a) is the amount that an object's image is enlarged.
   (b) is the extent to which detail in an image is preserved during the magnifying process.
   (c) is the degree to which image details stand out against their background.
   (d) focuses light rays emanating from an object to produce an image.

_____ 2. Resolving power
   (a) is the amount that an object's image is enlarged.
   (b) is the extent to which detail in an image is preserved during the magnifying process.
   (c) is the degree to which image details stand out against their background.
   (d) focuses light rays emanating from an object to produce an image.

_____ 3. A lens
   (a) is the amount that an object's image is enlarged.
   (b) is the extent to which detail in an image is preserved during the magnifying process
   (c) is the degree to which image details stand out against their background.
   (d) focuses light rays emanating from an object to produce an image.

_____ 4. Contrast
   (a) is the amount that an object's image is enlarged.
   (b) is the extent to which detail in an image is preserved during the magnifying process.
   (c) is the degree to which image details stand out against their background.
   (d) focuses light rays emanating from an object to produce an image.

_____ 5. The maximum useful magnification for a light microscope is about
   (a) $100\times$.
   (b) $1000\times$.
   (c) $10,000\times$.
   (d) $100,000\times$.

_____ 6. The two image-forming lenses of a compound light microscope are
   (a) the condenser and objective.
   (b) the condenser and ocular.
   (c) the objective and ocular.
   (d) none of these choices

_____ 7. Dyes are usually added to sections of biological specimens to increase
   (a) resolving power.
   (b) magnification.
   (c) contrast.
   (d) all of the above

_____ 8. If the magnification of the two image-forming lenses are both $10\times$, the total magnification of the image will be
   (a) $1\times$.
   (b) $10\times$.
   (c) $100\times$.
   (d) $1000\times$.

_____ 9. The distance through which a microscopic specimen can be moved and still have it remain in focus is called the
   (a) field of view.
   (b) working distance.
   (c) depth of field.
   (d) magnification.

_____ 10. Electron microscopes differ from light microscopes in that
   (a) electrons are used instead of light.
   (b) magnetic lenses replace glass lenses.
   (c) the electron path has to be maintained in a high vacuum.
   (d) a, b, and c are all true.

EXERCISE 3

Microscopy

Post-Lab Questions

3.1 Compound Light Microscope

1. What is the function of the following parts of a compound light microscope?
   (a) condenser lens

   FOCUSES lIGHT SOURCE

   (b) iris diaphragm

   CONTROls AMOUNT of lIGHT SOURCE

   (c) objective

   (d) ocular

2. In order, list the lenses in the light path between a specimen viewed with the compound light microscope and its image on the retina of the eye.

3. What happens to contrast and resolving power when the aperture of the condenser (i.e., the size of the hole through which light passes before it reaches the specimen) of a compound light microscope is decreased?

4. What happens to the field of view in a compound light microscope when the total magnification is increased?

5. Describe the importance of the following concepts to microscopy.
   (a) magnification

   (b) resolving power

   (c) contrast

6. Which photomicrograph of unstained cotton fibers was taken with the iris diaphragm closed? ___

A

B

7. Describe how you would care for and put away your compound light microscope at the end of the lab.

TURN RHEOSTAT All THE WAY DOWN
PUT STAGE All THE WAY DOWN
REVOLVE NOSEPIECE TO lOWEST hCAN POWER (4x)
TURN OFF
COVER W/ DUST-COVER.

3.2 How to Make a Wet Mount

8. Describe how to make a wet mount.

3.5 Other Microscopes

9. A camera mounted on a _____ microscope took this photo of a cut piece of cork.

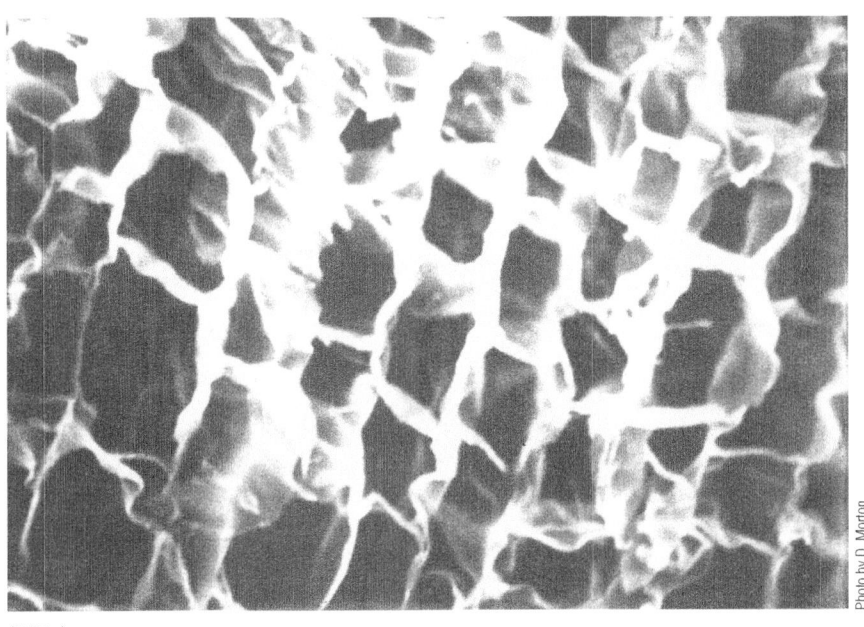

(515x)

Food for Thought

10. Why were humans unaware of microorganisms for most of their history?

Journi Norman

Diffusion, Osmosis, and the Functional Significance of Biological Membranes

OBJECTIVES

After completing this exercise, you will be able to

1. define *solvent, solute, solution, selectively permeable, diffusion, osmosis, concentration gradient, equilibrium, turgid, plasmolyzed, plasmolysis, turgor pressure, tonicity, hypertonic, isotonic, hypotonic;*

2. describe the structure of cellular membranes;

3. distinguish between diffusion and osmosis;

4. determine the effects of concentration and temperature on diffusion;

5. describe the effects of hypertonic, isotonic, and hypotonic solutions on red blood cells and *Elodea* leaf cells.

INTRODUCTION

Water is a great environment. Earthly life is believed to have originated in the water. Without it, life as we know it would cease to exist. Recently, the discovery of water in meteorites originating within our solar system has fueled speculation that life may not be unique to earth.

Living cells are made up of 75–85% water. Virtually all substances entering and leaving cells are dissolved in water, making it the **solvent** most important for life processes. The substances dissolved in water are called **solutes** and include such substances as salts and sugars. The combination of a solvent and dissolved solute is a **solution.** The cytoplasm of living cells contains numerous solutes, like sugars and salts, in solution.

All cells possess membranes composed of a phospholipid bilayer that contains different kinds of embedded and surface proteins. Look at Figure 4-1 to get an idea of the complexity of a cellular membrane.

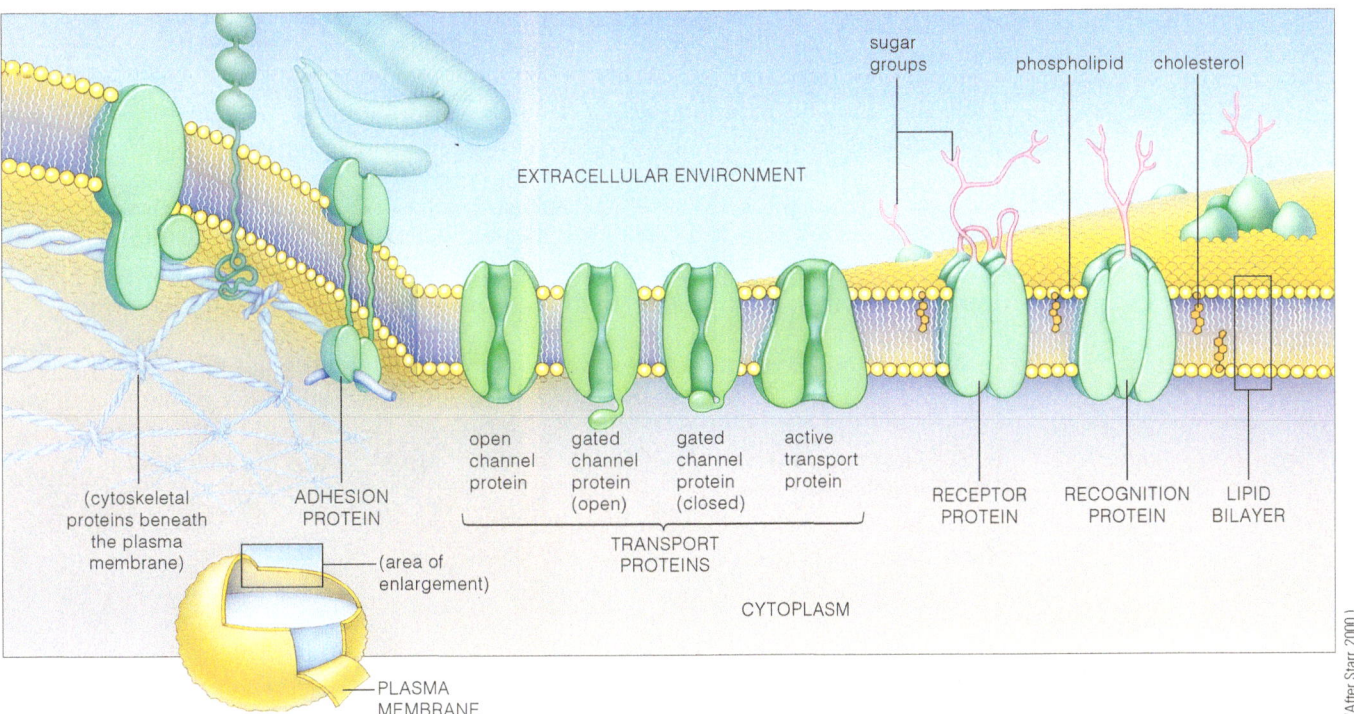

Figure 4-1 Artistic rendering of cutaway view of part of the plasma membrane.

Membranes are boundaries that solutes must cross to reach the cellular site where they will be utilized in the processes of life. These membranes regulate the passage of substances into and out of the cell. They are **selectively permeable,** allowing some substances to move easily while completely excluding others.

The most simple means by which solutes enter the cell is **diffusion,** the movement of solute molecules from a region of high concentration to one of lower concentration. Diffusion occurs without the expenditure of cellular energy. Once inside the cell, solutes move through the cytoplasm by diffusion, sometimes assisted by cytoplasmic streaming.

Water (the solvent) also moves across the membrane. **Osmosis** is the movement of *water* across selectively permeable membranes. Think of osmosis as a special form of diffusion, one occurring from a region of higher *water* concentration to one of lower *water* concentration.

The difference in concentration of like molecules in two regions is called a **concentration gradient.** Diffusion and osmosis take place *down* concentration gradients. Over time, the concentration of solvent and solute molecules becomes equally distributed, the gradient ceasing to exist. At this point, the system is said to be at **equilibrium.**

Molecules are always in motion, even at equilibrium. Thus, solvent and solute molecules continue to move because of randomly colliding molecules. However, at equilibrium there is no *net change* in their concentrations.

This exercise introduces you to the principles of diffusion and osmosis.

Note: **If Sections 4.2 and 4.3 are to be done during this lab period, start them before doing any other activity in this exercise.**

4.1 Experiment: Rate of Diffusion of Solutes *(About 10 min.)*

Solutes move within a cell's cytoplasm largely because of diffusion. However, the rate of diffusion (the distance diffused in a given amount of time) is affected by such factors as temperature and the size of the solute molecules. In this experiment, you will discover the effects of these two factors in gelatin (the substance of Jell-O®), a substance much like cytoplasm and used to simulate it in this experiment.

MATERIALS

Per student:

- metric ruler

Per student group (table):

- 1 set of 3 screw-cap test tubes, in rack, each half-filled with 5% gelatin, to which the following dyes have been added: potassium dichromate, aniline blue, Janus green; labeled with each dye and marked "5°C"

- 1 set of 3 screw-cap test tubes, in rack, as above but marked "Room Temperature"

Per lab room:

- 5°C refrigerator

PROCEDURE

Two sets of three screw-cap test tubes have been half-filled with 5% gelatin; and 1 mL of a dye has been added to each test tube. Set 1 is in a 5°C refrigerator; set 2 is at room temperature. Record the time at which your instructor tells you the experiment was started: _____

1. Remove set 1 from the refrigerator and compare the distance the dye has diffused in corresponding tubes of each set.

Caution

Be certain the cap to each tube is tight!

2. Invert and hold each tube vertically in front of a white sheet of paper. Use a metric ruler to measure how far each dye has diffused from the gelatin's surface. Record this distance in Table 4-1.

3. Determine the *rate* of diffusion for each dye by using the following formula:

rate of diffusion = distance ÷ elapsed time (hours)

Time experiment ended: _____

Time experiment started: _____

Elapsed time: _____ hours

Solute (dye)	Set 1 (5°C)		Set 2 (Room Temp.)	
	Distance (mm)	Rate	Distance (mm)	Rate
Potassium dichromate (MW = 294)[a]				
Janus green (MW = 511)				
Aniline blue (MW = 738)				

[a]MW = molecular weight, a reflection of the mass of a substance. To determine MW, add the atomic weights of all elements in a compound.

Which of the solutes diffused the slowest (regardless of temperature)? _____

Which diffused the fastest? _____

What effect did temperature have on the rate of diffusion? _____

Make a conclusion about the diffusion of a solute in a gel, relating the rate of diffusion to the molecular weight of the solute and to temperature.

_____

_____

Note: Return set 1 to the refrigerator.

4.2   Experiment: Osmosis *(About 20 min. for setup)*

Osmosis occurs when different concentrations of water are separated by a selectively permeable membrane. One example of a selectively permeable membrane within a living cell is the plasma membrane. In this experiment, you will learn about osmosis using dialysis membrane, a selectively permeable cellulose sheet that permits the passage of water but obstructs passage of larger molecules. If you examined the membrane with a scanning electron microscope, you would see that it is porous; it thus prevents molecules larger than the pores from passing through the membrane.

MATERIALS

Per student group (4):

- four 15-cm lengths of dialysis tubing, soaking in dH$_2$O
- eight 10-cm pieces of string or waxed dental floss
- ring stand and funnel apparatus (Figure 4-2)
- 25-mL graduated cylinder
- 4 small string tags
- china marker
- four 400-mL beakers

Per student group (table):

- dishpan half-filled with dH$_2$O
- paper toweling
- balance

Per lab room:

- source of dH$_2$O (at each sink)
- 15% and 30% sucrose solutions
- scissors (at each sink)

PROCEDURE

Work in groups of four for this experiment.

1. Obtain four sections of dialysis tubing, each 15 cm long, that have been presoaked in dH$_2$O. Recall that the dialysis tubing is permeable to water molecules but not to sucrose.
2. Fold over one end of each tube and tie it tightly with string or dental floss.
3. Attach a string tag to the tied end of each bag and number them 1–4.

4. Slip the open end of the bag over the stem of a funnel (Figure 4-2). Using a graduated cylinder* to measure volume, fill the bags as follows:

Bag 1. 10 mL of dH_2O
Bag 2. 10 mL of 15% sucrose

Bag 3. 10 mL of 30% sucrose
Bag 4. 10 mL of dH_2O

5. As each bag is filled, force out excess air by squeezing the bottom end of the tube.
6. Fold the end of the bag and tie it securely with another piece of string or dental floss.
7. Rinse each filled bag in the dishpan containing dH_2O; gently blot off the excess water with paper toweling.
8. Weigh each bag to the nearest 0.5 g.
9. Record the weights in the column marked "0 min." in Table 4-2.
10. Number four 400-mL beakers with a china marker.
11. Add 200 mL of dH_2O to beakers 1–3.
12. Add 200 mL of 30% sucrose solution to beaker 4.
13. Place bags 1–3 in the correspondingly numbered beakers.
14. Place bag 4 in the beaker containing 30% sucrose.
15. After 15 minutes, remove each bag from its beaker, blot off the excess fluid, and weigh each bag.
16. Record the weight of each bag in Table 4-2.
17. Return the bags to their respective beakers immediately after weighing.
18. Repeat steps 15–17 at 30, 45, and 60 minutes from time zero.

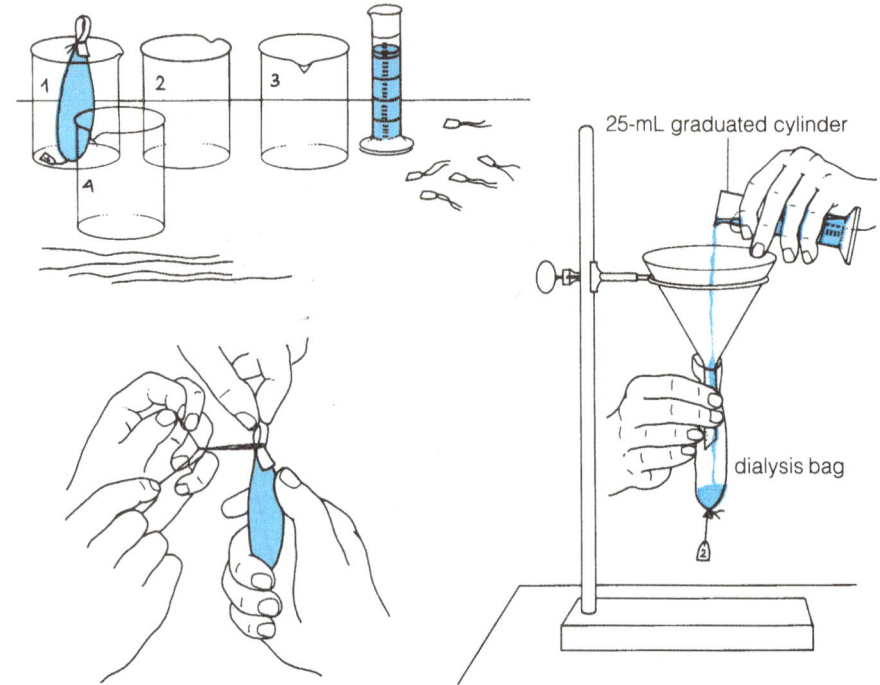

Figure 4-2 Method for filling dialysis bags.

At the end of the experiment, take the bags to the sink, cut them open, pour the contents down the drain, and discard the bags in the wastebasket. Pour the contents of the beakers down the drain and wash them according to the instructions given on page x.

Make a *qualitative* statement about what you have observed. WEIGHT SHIFT. DIFFERENT SOLUTIONS. DIFFERENT COLOR STRINGS.

Was the direction of *net* movement of water in bags 2–4 into or out of the bags? OUT OF the BAGS.

Which bag gained the most weight? Why? BAG 2, BECAUSE BEAKER CONTAINED SUCROSE.

*Be sure to rinse the cylinder if it has been used to measure sucrose.

LAB REPORT #2 – April 3

TABLE 4-2 Change in Weight as a Consequence of Osmosis

Bag	Bag Contents	Beaker Contents	Bag Weight (g)					Weight Change (g)
			0 min.	10 min.	20 min.	30 min.	min.	
1 *pink*	dH$_2$O	dH$_2$O	10.7	10.5	10	10.2		0.5 g
2 *green*	15% sucrose	dH$_2$O	10.3	11.3	10.9	11.4		1.1 g
3 *black*	30% sucrose	dH$_2$O	11	11.5	11.6	11.9		0.9 g
4 *red*	dH$_2$O	30% sucrose	10.4	9.4	8.7	7.9		2.5 g

4.3 Experiment: Selective Permeability of Membranes (*About 15 min. for setup*)

Dialysis tubing is a selectively permeable material that provides a means to demonstrate the movement of substances through cellular membranes.

MATERIALS

Per student group (4):

- 1 25-cm length of dialysis tubing, soaking in dH$_2$O
- two 10-cm pieces of string or waxed dental floss
- bottle of 1% soluble starch in 1% sodium sulfate (Na$_2$SO$_4$)
- dishpan half-filled with dH$_2$O
- 400-mL graduated beaker
- ring stand and funnel apparatus (Figure 4-2)
- bottle of 1% albumin in 1% sodium chloride (NaCl)
- 8 test tubes
- test tube rack
- china marker

- 25-mL graduated cylinder
- iodine (I$_2$KI) solution in dropping bottle
- 2% barium chloride (BaCl$_2$) in dropping bottle
- 2% silver nitrate (AgNO$_3$) in dropping bottle
- Biuret reagent in dropping bottle
- albustix reagent strips (optional)
- scissors

Per lab room:

- series of 4 test tubes in test tube rack demonstrating positive tests for starch, sulfate ion, chloride ion, protein

PROCEDURE

Work in groups of four.

1. Obtain a 25-cm section of dialysis tubing that has been soaked in dH$_2$O.
2. Fold over one end of the tubing and tie it securely with string or dental floss to form a leakproof bag (Figure 4-2).
3. Slip the open end of the bag over the stem of a funnel and fill the bag approximately half full with 25 mL of a solution of 1% soluble starch in 1% sodium sulfate (Na$_2$SO$_4$).
4. Remove the bag from the funnel; fold and tie the open end of the bag.
5. Rinse the tied bag in a dishpan partially filled with dH$_2$O.
6. Pour 200 mL of a solution of 1% albumin (a protein) in 1% sodium chloride (NaCl) into a 400-mL beaker.
7. Place the bag into the fluid in the beaker.
8. Record the time: __12:45 PM__
9. With a china marker, label eight test tubes, numbering them 1–8.
10. Seventy-five minutes after the start of the experiment, pour 20 mL of the *beaker contents* into a *clean* 25-mL graduated cylinder.
11. Decant (pour out) 5 mL from the graduated cylinder into each of the first four test tubes.

BARIUM CHLORIDE TO TEST for SULFATE (SO$_4$)

SILVER NITRATE TO TEST for CHLORIDE

12. Perform the following tests, recording your results in Table 4-3. Your instructor will have a series of test tubes showing positive tests for starch, sulfate and chloride ions, and proteins. You should compare your results with the known positives.

   (a) *Test for starch.* Add several drops of iodine solution (I_2KI) from the dropper bottle to test tube 1. If starch is present, the solution will turn blue-black.

   (b) *Test for sulfate ion.* Add several drops of 2% barium chloride ($BaCl_2$) from the dropper bottle to test tube 2. If sulfate ions (SO^{-4}) are present, a white precipitate of barium sulfate ($BaSO_4$) will form.

   (c) *Test for chloride ion.* Add several drops of 2% silver nitrate ($AgNO_3$) from the dropper bottle to test tube 3. A milky-white precipitate of silver chloride ($AgCl$) indicates the presence of chloride ions (Cl^2).

   (d) *Test for protein.* Add several drops of Biuret reagent from the dropper bottle to test tube 4. If protein is present, the solution will change from blue to pinkish-violet. The more intense the violet hue, the greater the quantity of the protein.

   An alternative method for determining the presence of protein is the use of albustix reagent strips. Presence of protein is indicated by green or blue-green coloration of the paper.

13. Wash the graduated cylinder, using the technique described on page x.
14. Thoroughly rinse the bag in the dishpan of dH_2O.
15. Using scissors, cut the bag open and empty the contents into the 25-mL graduated cylinder.
16. Decant 5-mL samples into each of the four remaining test tubes.
17. Perform the tests for starch, sulfate ions, chloride ions, and protein on tubes 5–8, respectively.
18. Record the results of this series of tests in Table 4-4.

To which substances was the dialysis tubing permeable?
dH₂O   SUCROSE

What physical property of the dialysis tubing might explain its differential permeability?
THIN, CLEAR, OPENED ON BOTH ENDS,

19. Discard contents of test tubes and beaker down sink drain. Wash glassware by using the technique described on page x.
20. Discard dialysis tubing in wastebasket.

TABLE 4-3 Results of Tests for Substances in Beaker[a]

	At Start of Experiment		
Starch	−	+	
Sulfate ion	−	+	
Chloride ion	+	+	
Albumin	+	−	

[a]Contents of beaker: (+) = presence, (−) = absence.

+ CHLORIDE ION = CLOUDY

TABLE 4-4 Results of Tests for Substances in Dialysis Bag[a]

	At Start of Experiment		
Starch	+	+	
Sulfate ion	+	+	
Chloride ion	−	−	
Albumin	−	+	

[a]Contents of dialysis bag: (+) = presence, (−) = absence.

Plant cells are surrounded by a rigid cell wall, composed primarily of the glucose polymer, cellulose. Recall from Exercise 6 that many plant cells have a large central vacuole surrounded by the vacuolar membrane. The vacuolar membrane is selectively permeable. Normally, the solute concentration within the cell's central vacuole is greater than that of the external environment. Consequently, water moves into the cell, creating **turgor pressure**, which presses the cytoplasm against the cell wall. Such cells are said to be **turgid.** Many nonwoody plants (like beans and peas) rely on turgor pressure to maintain their rigidity and erect stance.

In this experiment, you will discover the effect of external solute concentration on the structure of plant cells.

MATERIALS

Per student:

- forceps
- 2 microscope slides
- 2 coverslips
- compound microscope

Per student group (table):

- *Elodea* in tap water
- 2 dropping bottles of dH_2O
- 2 dropping bottles of 20% sodium chloride (NaCl)

PROCEDURE

1. With a forceps, remove two young leaves from the tip of an *Elodea* plant.
2. Mount one leaf in a drop of distilled water on a microscope slide and the other in 20% NaCl solution on a second microscope slide.
3. Place coverslips over both leaves.
4. Observe the leaf in distilled water with the compound microscope. Focus first with the medium-power objective and then switch to the high-dry objective.
5. Label the photomicrograph of turgid cells (Figure 4-3).
6. Now observe the leaf mounted in 20% NaCl solution. After several minutes, the cell will have lost water, causing it to become **plasmolyzed.** (This process is called **plasmolysis.**) Label the plasmolyzed cells shown in Figure 4-4.

Figure 4-3   Turgid *Elodea* cells (400×). (Photo by J. W. Perry.)
Labels: cell wall, chloroplasts in cytoplasm, central vacuole

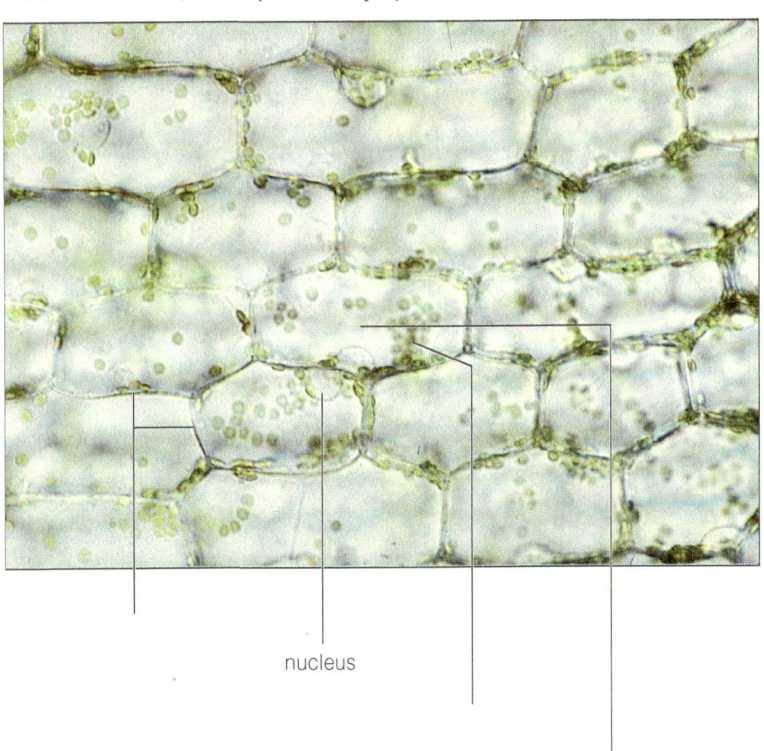

nucleus

Tonicity describes one solution's solute concentration compared to that of another solution. The solution containing the lower concentration of solute molecules than another is **hypotonic** *relative to the second solution.* Solutions containing equal concentrations of solute are **isotonic** to each other, while one containing a greater concentration of solute relative to a second one is **hypertonic.**

Were the contents of the vacuole in the *Elodea* leaf in distilled water ⟨hypotonic,⟩ isotonic, or hypertonic compared to the dH_2O? __*hypotonic*__

Was the 20% NaCl solution ⟨hypertonic,⟩ isotonic, or hypotonic relative to the cytoplasm? _____
__*hypertonic*__

If a hypotonic and a hypertonic solution are separated by a selectively permeable membrane, in which direction will the water move? __*hypo [Hi] → [Low]*__

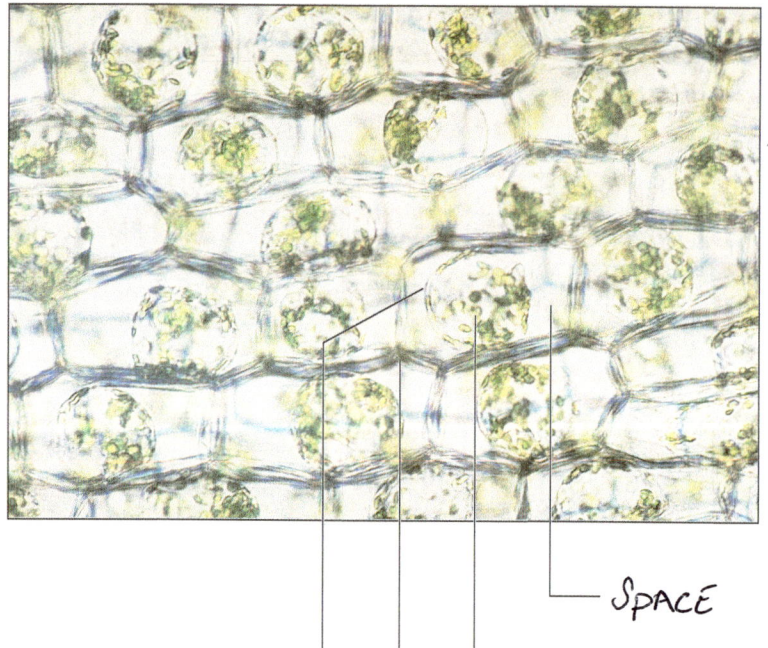

Figure 4-4 Plasmolyzed *Elodea* cells (400×). (Photo by J. W. Perry.)
Labels: ~~green chloroplasts~~ in cytoplasm, ~~plasma membrane~~, ~~space~~ (between cell wall and plasma membrane)

SPACE

CHLOROPLASTS

CELL WALL

PLASMA MEMBRANE

Name two selectively permeable membranes that are present within the *Elodea* cells and that were involved in the plasmolysis process.

1. PLASMA MEMBRANE
2. CYTOPLASM

| 4.5 | **Experiment: Osmotic Changes in Red Blood Cells** *(About 15 min.)* |

Animal cells lack the rigid cell wall of a plant. The external boundary of an animal cell is the selectively permeable plasma membrane. Consequently, an animal cell increases in size as water enters the cell. However, since the plasma membrane is relatively fragile, it ruptures when too much water enters the cell. This is because of excessive pressure pushing out against the membrane. Conversely, if water moves out of the cell, it becomes plasmolyzed and looks spiny.

In this experiment, you will use red blood cells to discover the effects of osmosis in animal cells.

MATERIALS

Per student:

- compound microscope

Per student group (4):

- 3 clean screw-cap test tubes
- test tube rack
- metric ruler
- china marker
- bottle of 0.9% sodium chloride (NaCl)
- bottle of 10% NaCl
- bottle of dH$_2$O

- 3 disposable plastic pipets
- 3 clean microscope slides
- 3 coverslips

Per student group (table):

- bottle of sheep blood (in ice bath)

Per lab room:

- source of dH$_2$O

PROCEDURE

Work in groups of four for this experiment, but do the microscopic observations individually.

1. Observe the scanning electron micrographs in Figure 4-5.

Figure 4-5a illustrates the normal appearance of red blood cells. They are biconcave disks; that is, they are circular in outline with a depression in the center of both surfaces. Cells in an isotonic solution will appear like these blood cells.

Figure 4-5b shows cells that have been plasmolyzed. (In the case of red blood cells, plasmolysis is given a special term, *crenation;* the blood cell is said to be *crenate.*)

Figure 4-5c represents cells that have taken in water but have not yet burst. (Burst red blood cells are said to be *hemolyzed,* and of course they can't be seen.) Note their swollen, spherical appearance.

2. Obtain three clean screw-cap test tubes.
3. Lay test tubes 1 and 2 against a metric ruler and mark lines indicating 5 cm *from the bottom of each tube.*
4. Fill each tube as follows:
   Tube 1: 5 cm of 0.9% sodium chloride (NaCl)
            5 drops of sheep blood
   Tube 2: 5 cm of 10% NaCl
            5 drops of sheep blood
5. Lay test tube 3 against a metric ruler and mark lines indicating 0.5 cm and 5 cm *from the bottom of the tube.*
6. Fill tube 3 to the 0.5-cm mark with 0.9% NaCl, and to the 5-cm mark with dH_2O. Then add 5 drops of sheep blood. Enter the contents of each tube in the appropriate column of Table 4-5.

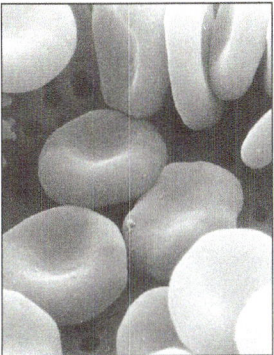

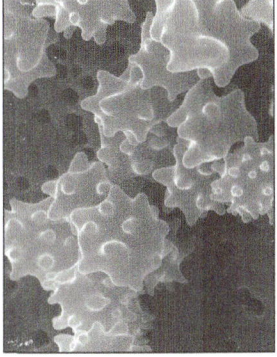

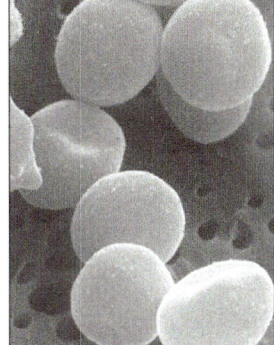

a Red blood cells in an isotonic solution ("normal")

b Red blood cells in a hypertonic solution ("crenate")

c Red blood cells in a hypotonic solution

Figure 4-5 Scanning electron micrographs of red blood cells. (Photos from M. Sheetz, R. Painter, and S. Singer. Reproduced from *The Journal of Cell Biology,* 1976, 70:193, by copyright permission of the Rockefeller University Press and M. Sheetz.)

7. Replace the caps and mix the contents of each tube by inverting several times (Figure 4-6a).
8. Hold each tube flat against the printed page of your lab manual (Figure 4-6b). *Only if the blood cells are hemolyzed should you be able to read the print.*
9. In Table 4-5, record your observations in the column "Print Visible?"
10. Number three clean microscope slides.
11. With three *separate* disposable pipets, remove a small amount of blood from each of the three tubes. Place 1 drop of blood from tube 1 on slide 1, 1 drop from tube 2 on slide 2, and 1 drop from tube 3 on slide 3.
12. Cover each drop of blood with a coverslip.
13. Observe the three slides with your compound microscope, focusing first with the medium-power objective and finally with the high-dry objective. (Hemolyzed cells are virtually unrecognizable; all that remains are membranous "ghosts," which are difficult to see with the microscope.)
14. In Figure 4-7, sketch the cells from each tube. Label the sketches, indicating whether the cells are normal, plasmolyzed (crenate), or hemolyzed.

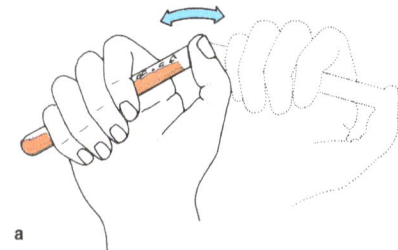

a

b

(After Abramoff and Thomson, 1982.)

Figure 4-6 Method for studying effects of different solute concentrations on red blood cells.

15. Record the microscopic appearance in Table 4-5.
16. Record the relative tonicity of the sodium chloride solutions you added to the test tubes in Table 4-5.

Why do red blood cells burst when put in a hypotonic solution whereas *Elodea* leaf cells do not?

_____

_____

_____

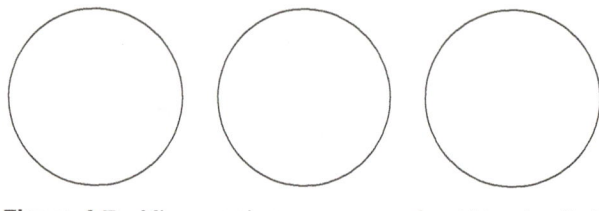

Figure 4-7 Microscopic appearance of red blood cells in different solute concentrations (_____×).
Labels: normal, plasmolyzed (crenate), hemolyzed

After completing all experiments, take your dirty glassware to the sink and wash it as directed on page x. Invert the test tubes in the test tube rack so they drain. Reorganize your work area, making certain all materials used in this exercise are present for the next class.

TABLE 4-5 Effect of Salt Solutions on Red Blood Cells

Tube	Contents	Print Visible?	Microscopic Appearance of Cells	Tonicity of External Solution[a]
1				
2				
3				

[a]With respect to that inside the red blood cell at the start of the experiment.

4.6 Experiment: Determining the Concentration of Solutes in Cells
(About 20 min. for setup)

If you've done the previous experiments of this exercise, you now know that water flows into or out of cells in response to the concentration of solutes within the cells. But you might logically ask at this point how much solute is present in a typical cell. While the answer varies from cell to cell, a simple experiment enables you to determine the osmotic concentration in the cells of a potato tuber.

MATERIALS

Per student group (4):

- five 250-mL beakers
- large potato tuber
- china marker
- single-edge razor blades *or* paring knife
- metric ruler
- potato peeler

Per table:

- balance
- paper toweling
- bottles containing solutions of 0.15, 0.20, 0.25, 0.30, and 0.35 M sucrose

PROCEDURE

1. With the china marker, label the five 250-mL beakers with the concentrations of sucrose solution.
2. Pour about 100 mL of each solution into its respective beaker.
3. Peel the potato and then cut it into five 3-cm cubes (3 cm on each side).
4. Without delay, weigh each cube to the nearest 0.01 g. Record the weights in Table 4-6.
5. Place one cube in each beaker and allow it to remain there for a minimum of 30 minutes, longer if time is available.

6. After the experimental period has elapsed, remove each cube, one at a time, and blot it lightly but thoroughly with the paper toweling.
7. Weigh each cube and record its final weight in Table 4-6. Then calculate and record the weight loss or gain.
8. Calculate the percent change in weight by dividing the *initial weight* by the *final weight*.

	Weight			
Solution	Initial	Final	Change	Percent Change
0.15 M				
0.20 M				
0.25 M				
0.30 M				
0.35 M				

TABLE 4-6 Determining the Solute Concentration in Potato Tuber Cells

The cube with the lowest percentage of weight loss or gain is in a solution that most closely approximates the solute concentration of the cells within the potato tuber. Of course, most of the solute within the tuber is in the form of starch, and our experimental solution is sucrose. The results of this experiment indicate that the *concentration* of the solute, but not the *type* of solute, is important for osmosis to occur.

What was the approximate concentration of solute in the potato tuber? _____

Which concentration resulted in the *greatest* percentage change? _____

Make a statement that relates the amount of water loss or gain to the concentration of the solute. _____

_____

_____

_____

C 1. If one were to identify the most important compound for sustenance of life, it would probably be
(a) salt
(b) $BaCl_2$
(c) water
(d) I_2KI

A 2. A solvent is
(a) the substance in which solutes are dissolved
(b) a salt or sugar
(c) one component of a biological membrane
(d) selectively permeable

B 3. Diffusion
(a) is a process requiring cellular energy
(b) is the movement of molecules from a region of higher concentration to one of lower concentration
(c) occurs only across selectively permeable membranes
(d) ~~none of the above~~

D 4. Cellular membranes
(a) consist of a phospholipid bilayer containing embedded proteins
(b) control the movement of substances into and out of cells
(c) are selectively permeable
(d) are all of the above

C 5. An example of a solute would be
(a) Janus green B
(b) water
(c) sucrose
(d) both a and c

A 6. Dialysis membrane is
(a) selectively permeable
(b) used in these experiments to simulate cellular membranes
(c) permeable to water but not to sucrose
(d) all of the above

C 7. Specifically, osmosis
(a) requires the expenditure of cellular energy
(b) is diffusion of water from one region to another
(c) is diffusion of water across a selectively permeable membrane
(d) is none of the above

_____ 8. Which of the following reagents does *not* fit with the substance being tested for?
(a) Biuret reagent protein
(b) $BaCl_2$ starch
(c) $AgNO_3$ chloride ion
(d) albustix protein

A 9. When the cytoplasm of a plant cell is pressed against the cell wall, the cell is said to be
(a) turgid
(b) plasmolyzed
(c) hemolyzed
(d) crenate

C 10. If one solution contains 10% NaCl and another contains 30% NaCl, the 30% solution is
(a) isotonic
(b) hypotonic
(c) hypertonic
(d) plasmolyzed, with respect to the 10% solution

*Dissolved - solute
salt, sugar

EXERCISE 4

Diffusion, Osmosis, and the Functional Significance of Biological Membranes

POST-LAB QUESTIONS

4.1 Experiment: Rate of Diffusion of Solutes

1. You want to dissolve a solute in water. Without shaking or swirling the solution, what might you do to increase the rate at which the solute would go into solution? Relate your answer to your method's effect on the motion of the molecules.

4.2 Experiment: Osmosis

2. If a 10% sugar solution is separated from a 20% sugar solution by a selectively permeable membrane, in which direction will there be a net movement of water?

3. Based on your observations in this exercise, would you expect dialysis membrane to be permeable to sucrose? Why?

4.4 Experiment: Plasmolysis in Plant Cells

4. You are having a party and you plan to serve celery, but your celery has gone limp, and the stores are closed. What might you do to make the celery crisp (turgid) again?

5. Why don't plant cells undergo osmotic lysis?

6. This drawing represents a plant cell that has been placed in a solution.
   a. What *process* is taking place in the direction of the arrows? What is happening at the cellular level when a wilted plant is watered and begins to recover from the wilt?

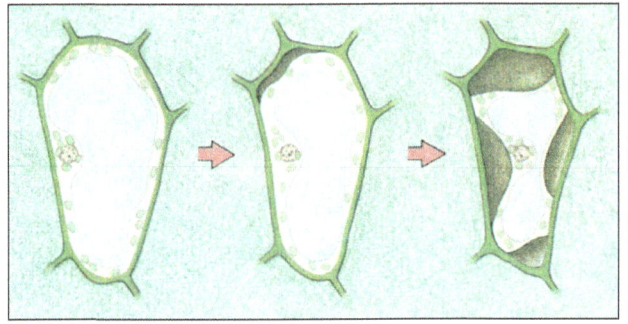

(After Starr and Taggart, 1989.)

   b. Is the solution in which the cells have been placed hypotonic, isotonic, or hypertonic relative to the cytoplasm?

4.5 Experiment: Osmotic Changes in Red Blood Cells

7. A human lost at sea without fresh drinking water is effectively lost in an osmotic desert. Why would drinking salt water be harmful?

Food for Thought

8. How does diffusion differ from osmosis?

9. Plant fertilizer consists of numerous different solutes. A small dose of fertilizer can enhance plant growth, but overfertilization can kill the plant. Why might overfertilization have this effect?

10. What does the word *lysis* mean? (*Now* does the name of the disinfectant Lysol® make sense?)

Enzymes: Catalysts of Life

OBJECTIVES

After completing this exercise, you will be able to

1. define *catalyst, enzyme, activation energy, enzyme–substrate complex, substrate, product, active site, denaturation, cofactor*;

2. explain how an enzyme operates;

3. recognize benzoquinone as a brown substance formed in damaged plant tissue;

4. indicate the substrates for the enzyme catechol oxidase;

5. describe the effect of temperature on the rate of chemical reactions in general and on enzymatically controlled reactions in particular;

6. describe the effect that an atypical pH may have on enzyme action;

7. indicate how a cofactor might operate and identify a cofactor for catechol oxidase.

INTRODUCTION

Life as we know it is impossible without enzymes. The energy required by your muscles simply to open your lab manual would take years to accumulate without enzymes. Due to the presence of enzymes, the myriad chemical reactions occurring in your cells at this very moment are being completed in a fraction of a second rather than the years or even decades that would be otherwise required.

Enzymes are proteins that function as biological catalysts. A **catalyst** is a substance that lowers the amount of energy necessary for a chemical reaction to proceed. You might think of this so-called **activation energy** as a mountain to be climbed. Enzymes decrease the size of the mountain, in effect turning it into a molehill (Figure 5-1).

By lowering the activation energy, an enzyme affects the *rate* at which reaction occurs. Enzyme-boosted reactions may proceed from 100,000 to 10 million times faster than they would without the enzyme.

In an enzyme-catalyzed reaction, the reactant (the substance being acted upon) is called the **substrate.** Substrate molecules combine with enzyme molecules to form a temporary **enzyme–substrate complex. Products** are

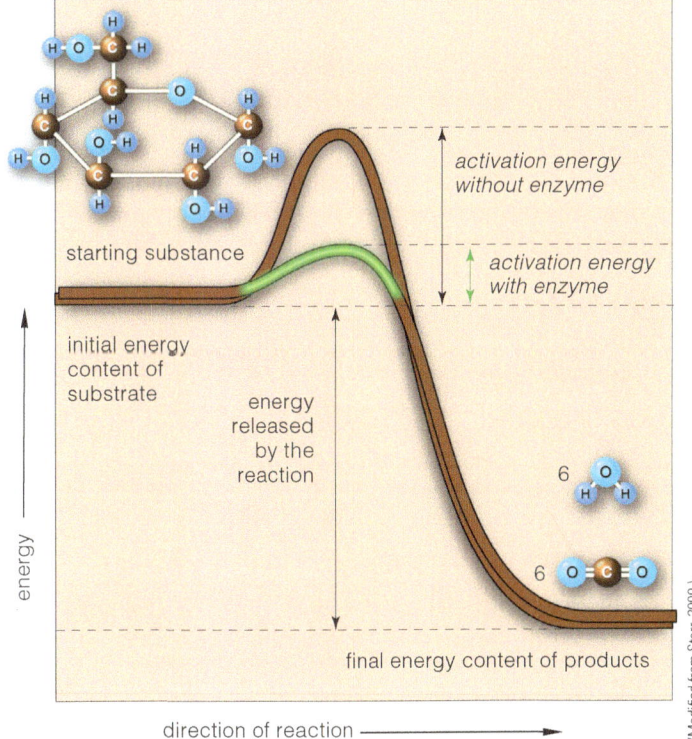

starting substance

initial energy content of substrate

energy released by the reaction

activation energy without enzyme

activation energy with enzyme

final energy content of products

energy

direction of reaction

(Modified from Starr, 2000.)

Figure 5-1 Enzymes and activation energy.

formed, and the enzyme molecule is released unchanged. Thus, the enzyme is not used up in the process and is capable of catalyzing the same reaction again and again. This can be summarized as follows:

substrate $\xrightarrow{\text{enzyme}}$ enzyme–substrate complex $\longrightarrow$ products + enzyme

Before we proceed, let's visualize an enzyme. Look at Figure 5-2. Using common, everyday items, think of an enzyme as a key that unlocks a lock. Imagine that you have a number of locks, and all use the same key. You would only need one key to unlock them.

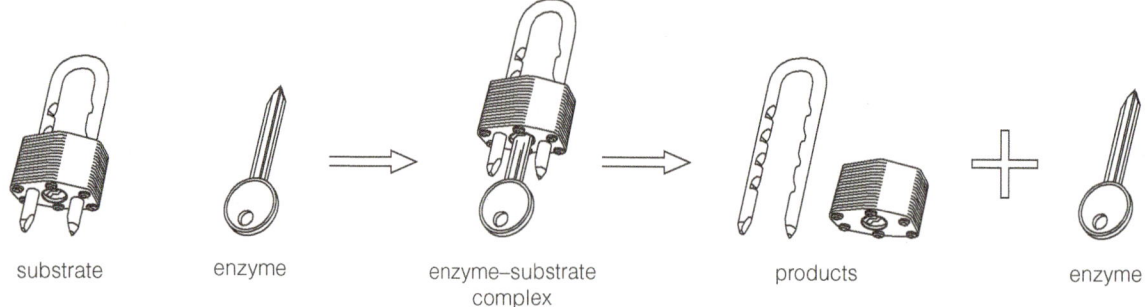

| substrate | enzyme | enzyme–substrate complex | products | enzyme |

Figure 5-2 Action of an enzyme.

Changing the shape of the key just a tiny bit may still allow the key to function in the lock, but you may have to fumble with the key a bit to get the lock open. Changing the key more results in an inability to open the lock. Similarly, a change in the shape of an enzyme alters its function. We will examine a number of factors in this exercise to determine their effects on enzyme action, including

1. temperature
2. pH (hydrogen-ion concentration of the environment)
3. specificity (how discriminating the enzyme is in catalyzing different potential substrates)
4. cofactor necessity (the need for a metallic ion for enzyme activity)

Although thousands of enzymes are present within cells, we'll examine only one: catechol oxidase (also known as tyrosinase).

Work in groups of four for all sections in this exercise.

5.1   Using a Spectronic 20 (Spec 20) to Determine Color Changes
(About 15 min.)

If you are not using a spectrophotometer, an instrument that measures color change, skip the following 11-step procedure. (See Appendix 2 for an explanation of how the Spec 20 works.)

MATERIALS

Per student:

- disposable plastic gloves

Per student group (4):

- ice bath with wash bottle of potato extract containing catechol oxidase

- bottle of dH$_2$O
- china marker
- Spec 20 spectrophotometer
- 1-mL pipet and bulb
- 5-mL pipet and bulb

PROCEDURE: ZEROING THE SPEC 20

1. Obtain a clean test tube designed to fit in the Spec 20.
2. With a wax pencil, label the top of it with a C for "control."
3. If the tube does not already have a vertical line on it, place one on it with the wax pencil. This mark must face the front of the sample holder.
4. Using the 1-mL pipet and bulb, measure 1 mL of potato extract into test tube C.
5. Using the 5-mL pipet and bulb, add 5 mL of dH$_2$O to the test tube.
6. Place your gloved thumb over the mouth of the test tube and invert the tube to mix the contents.
7. Adjust the wavelength knob (top, right) of the Spec 20 to 540 nm.
8. Rotate the lower *left* absorbance adjustment so the needle reads infinity (∞) on the bottom scale.
9. Clean the surface of the tube by wiping it with a tissue and insert the tube into the Spec 20 sample holder with the vertical mark facing front (toward you).
10. Rotate the lower *right* absorbance adjustment so the needle reads zero.
11. You have now zeroed the Spec 20 for this exercise. Remove tube C and set it aside.

The previous procedure is done to account for the fact that the potato extract has color and absorbs light. In your experiments, you will be measuring the change in the amount of light absorbed by potato extract before and after various experimental treatments.

<table>
<tr><td>**5.2**</td><td>**Formation and Detection of Benzoquinone** *(About 20 min.)*</td></tr>
</table>

Catechol oxidase is an enzyme that catalyzes the production of benzoquinone and water from catechol:

$$\text{catechol} + \tfrac{1}{2}\,O_2 \xrightarrow{\ \text{catechol oxidase}\ } \text{benzoquinone} + H_2O$$

(substrates)　(enzyme in potato extract)　(product)

This is an oxidation reaction, with catechol and oxygen as the substrates. Hence, the enzyme gets its preferred name, *catechol oxidase.* (This suffix *-ase* is a tipoff that the substance is an enzyme.)

Catechol and catechol oxidase are present in the cells of many plants, although in undamaged tissue they are separated in different compartments of the cells. Injury causes mixing of the substrate and enzyme, producing benzoquinone, a brown substance.

You've probably noticed the brown coloration of a damaged apple or the blackening of an injured potato tuber (Figure 5-3). Benzoquinone inhibits the growth of certain microorganisms that cause rot.

In this section, you will form the product, benzoquinone, and establish a color intensity scale or absorbance standard that you will use in subsequent experiments.

(Photo by J. W. Perry.)

Figure 5-3　Potato showing browning due to benzoquinone production. The potato on the left was sliced immediately before this photo was taken. The one on the right had been cut and exposed to the oxygen in the air for several minutes before being photographed.

MATERIALS

Per student:

- disposable plastic gloves

Per student group (4):

- 3 test tubes
- test tube rack
- metric ruler
- ice bath with wash bottle of potato extract containing catechol oxidase
- wash bottle containing 1% catechol solution

- bottle of dH_2O
- china marker
- warmed up and zeroed Spec 20, optional; see Appendix 2

Per lab room:

- 40°C waterbath
- vortex mixer (optional)

PROCEDURE

1. With a china marker, label three test tubes 2_A, 2_B, and 2_C. Place your initials on each test tube for later identification.
2. Lay the test tubes against a metric ruler and mark lines on the tubes corresponding to 1 cm and 2 cm *from the bottom* of each tube.
3. Fill each tube as follows:
   Tube 2_A:　1 cm of potato extract containing catechol oxidase
   　　　　　　1 cm of 1% catechol solution
   Tube 2_B:　1 cm of potato extract containing catechol oxidase
   　　　　　　1 cm of dH_2O
   Tube 2_C:　1 cm of 1% catechol solution
   　　　　　　1 cm of dH_2O

> **Caution**
>
> *Some of the chemicals (catechol, hydroquinone) used in these experiments can be hazardous to your health if they are ingested or taken in through your skin. Wear disposable plastic gloves for all experiments.*

4. Shake all tubes (using a vortex mixer if available).
5. Record the color of the solution in each tube in the "time 0" spaces of Table 5-1.
6. Place the tubes in a 40°C waterbath.
7. After 10 minutes, the catechol should be completely oxidized. Remove the tubes from the waterbath. Record the *color* of the substance in each test tube in Table 5-1.

   What you do next depends on whether you're using a spectrophotometer. Proceed with *either* step 8 *or* 9.

8. *Color-intensity method:* Consider the color of the product in tube 2_A to be a 5 on a color intensity scale of 0–5, while the color of the substance in tubes 2_B and 2_C is a 0. In Table 5-2, record the color intensity for tubes 2_B and 2_C as 0, and that of tube 2_A as 5.

   You will use this scale to make comparisons in Sections 5.3–5.6. Keep the contents in tubes 2_A, 2_B, and 2_C, and refer to them as you make comparisons in subsequent experiments.

9. *Spec 20 method:*
   (a) Pipet 6 mL of the contents of each tube into six separate Spec 20 tubes.
   (b) Clean off each tube.
   (c) Zero the Spec 20 using the contents of tube 2_C.
   (d) Insert tubes containing the contents of tube 2_A and then tube 2_B, determine the absorbance of each, and record them in Table 5-2.

TABLE 5-1   Formation and Detection of Benzoquinone: Record Color

Time	Tube 2_A: Potato Extract and Catechol	Tube 2_B: Potato Extract and Water	Tube 2_C: Catechol and Water
0 min.	CLEAR	CLEAR	CLEAR
10 min.	YELLOW	CLEAR	CLEAR

What is the brown-colored substance that appeared in tube 2_A? _BENZOQUINONE_

What is the substrate for the reaction that occurred in tube 2_A? _CATECHOL_

What is the product of the reaction in tube 2_A? _BENZOQUINONE_

What substances do tubes 2_B and 2_C lack that account for the absence of the brown-colored substance?

2_B _SUBSTRATE_ 2_C ~~DOWN~~ _ENZYME_

What is the purpose of tubes 2_B and 2_C? _SHOW VARIATION._

PROVE YOU NEED BOTH FOR REACTION

TABLE 5-2   Color-Intensity Scale or Absorbance

Intensity/Absorbance	Tube	Color of Product
0	2_C	CLEAR
0	2_B	CLEAR
2	2_A	YELLOW

5.3   Experiment: Enzyme Specificity *(About 20 min.)*

Generally, enzymes are substrate-specific, acting on one particular substrate or a small number of structurally similar substrates. This specificity is due to the three-dimensional structure of the enzyme. For the enzyme–substrate complex to form, the structure of the substrate must very closely complement that of the enzyme's

active site. The active site is a special region of the enzyme where the substrate binds. The active site has a small amount of moldability, so that the active site and substrate become fully complementary to each other, as shown in Figure 5-4.

Think again about the lock and key analogy (Figure 5-2). If the lock and key are not complementary, the lock won't open. But how exact a fit is necessary?

In this experiment, you will determine the ability of the enzyme catechol oxidase to catalyze the oxidation of two different but structurally similar substrates: catechol and hydroquinone. First, examine the chemical structure of each compound:

SUBSTRATE catechol   hydroquinone *All SUBSTRATE*

You need not memorize these structural formulas, but do notice that both are ring structures with two hydroxyl (—OH) groups attached.

Keep this in mind as you do the next experiment, in which you will determine how specific (discriminating) catechol oxidase is for particular substrates.

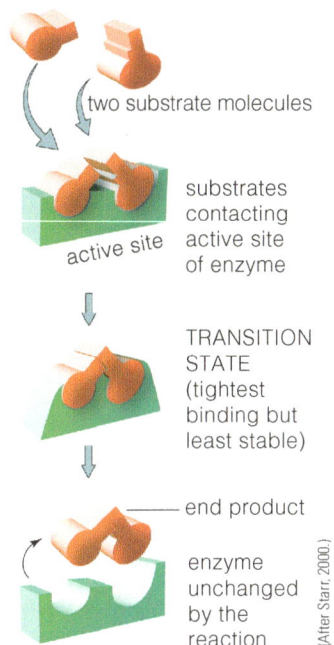

two substrate molecules

substrates contacting active site of enzyme

active site

TRANSITION STATE (tightest binding but least stable)

end product

enzyme unchanged by the reaction

(After Starr, 2000.)

Figure 5-4 Induced-fit model of enzyme–substrate interactions.

MATERIALS

Per student group (4):

- 3 test tubes
- test tube rack
- metric ruler
- china marker
- wash bottle containing 1% catechol
- wash bottle containing 1% hydroquinone
- ice bath with wash bottle of potato extract containing catechol oxidase
- warmed up and zeroed Spec 20, optional; see Appendix 6

Per lab room:

- 40°C waterbath
- vortex mixer (optional)

PROCEDURE

1. With a china marker, label three clean test tubes 3_A, 3_B, and 3_C. Include your initials for identification.
2. Lay the test tubes against a metric ruler and mark lines indicating 1 cm and 2 cm *from the bottom* of each test tube.
3. Fill each tube as follows:
    Tube 3_A: 1 cm of potato extract containing catechol oxidase
            1 cm of 1% catechol
    Tube 3_B: 1 cm of potato extract containing catechol oxidase
            1 cm of 1% hydroquinone
    Tube 3_C: 1 cm of potato extract containing catechol oxidase
            1 cm of dH_2O
4. Gently shake the test tubes to mix the contents.
5. Compare the color intensity of the solution in each test tube *with the standards produced in Section 5.2* and record them at time 0 in Table 5-3.
6. Place the test tubes in a 40°C waterbath.

This experiment addresses the hypothesis that *the structure of a substrate determines how well an enzyme acts upon the substrate.*

7. While you wait for the experiment to run its course, write your prediction of the outcome in Table 5-3.
8. After 10 minutes, remove the test tubes from the water bath and examine them. Choose step 9 *or* 10 and proceed.
9. *Color-intensity method:* Record the color intensity (scale 0–5) of each tube's contents in Table 5-3.

TABLE 5-3 Specificity of Catechol Oxidase for Different Substrates

Prediction:

| | Relative Color Intensity or Absorbance | | |
Time	Tube 3$_A$: Catechol	Tube 3$_B$: Hydroquinone	Tube 3$_C$: dH$_2$O
0 min.	CLEAR	CLEAR	CLEAR
10 min.	YELLOW	CLEAR	CLEAR

Conclusion:

10. *Spec 20 method:*
    (a) Pipet 6 mL of the contents of each tube into three labeled Spec 20 tubes.
    (b) Clean off each tube.
    (c) Zero the Spec 20 using the contents of tube 3$_C$.
    (d) Insert tubes containing the contents of tube 3$_A$ and then tube 3$_B$, determine the absorbance of each, and record them in Table 5-3.

11. Upon which substrate does catechol oxidase work best, forming the most benzoquinone in the shortest amount of time?

    3A

12. Based on your knowledge of the structures of the two substrates, what apparently determines the specificity of catechol oxidase?

    PLACEMENT of hydroxyl group

13. Why was tube 3$_C$ included in this experiment?

    AS CONTROLLED GROUP

14. Record your conclusion in Table 5-3, either accepting or rejecting the hypothesis.

5.4   **Experiment: Effect of Temperature on Enzyme Activity** *(25 min.)*

The rate at which chemical reactions take place is largely determined by the temperature of the environment. *Generally, for every 10°C rise in temperature, the reaction rate doubles.* Within a rather narrow range, this is true for enzymatic reactions also. However, because enzymes are proteins, excessive temperature alters their structure, destroying their ability to function. When an enzyme's structure is changed sufficiently to destroy its function, the enzyme is said to be **denatured.** Most enzymatically controlled reactions have an *optimum* temperature and pH—that is, one temperature and pH where activity is maximized.

   In this experiment, you will determine the temperature range over which the enzyme catechol oxidase is able to catalyze its substrate. You will also determine the best (optimum) temperature for the reaction.

MATERIALS

Per student group (4):

- 6 test tubes
- test tube rack
- metric ruler
- china marker
- wash bottle containing 1% catechol
- ice bath with wash bottle of potato extract containing catechol oxidase
- three 400-mL graduated beakers

- heat-resistant glove
- Celsius thermometer
- warmed up and zeroed Spec 20, optional; see Appendix 2

Per student group (table):

- hot plate *or* burner, tripod support, wire gauze, and matches or striker
- boiling chips

Per lab room:

- source of room-temperature water
- three waterbaths: 40°C, 60°C, 80°C
- vortex mixer (optional)

PROCEDURE

1. Half fill one 400-mL beaker with tap water. Add a few boiling chips and turn on the hotplate to the highest temperature setting, *or,* if your lab is equipped with burners, light the burner. Bring the water to a boil, then turn the heat down so that the water just continues to boil.
2. Put 150 mL of tap water into a second beaker and add ice to the water.
3. Half fill a third beaker with water from the source at room temperature.
4. With a china marker, label six test tubes 4_A–4_F. Include your initials for identification.
5. Lay the test tubes against a metric ruler and mark off lines indicating 1 cm and 2 cm *from the bottom* of each tube.
6. Fill each tube to the 1-cm mark with potato extract containing catechol oxidase.
7. **(a)** Place tube 4_A in the 400-mL beaker of ice water.
   Measure and record the water temperature: _____°C
   (b) Place tube 4_B in the 400-mL beaker containing room-temperature water.
   Room temperature: _____°C
   (c) Place tube 4_C in the 40°C waterbath.
   (d) Place tube 4_D in the 60°C waterbath.
   (e) Place tube 4_E in the 80°C waterbath.
   (f) Place tube 4_F in the 400-mL beaker containing boiling water.
   Temperature of boiling water: _____°C
8. Allow the test tubes to remain at the various temperatures for 5 minutes.
9. Remove the tubes and add catechol to the 2-cm line on each. Agitate the tubes (with a vortex mixer if available) to mix the contents.
10. *Color-intensity method:* In Table 5-4, record the relative color intensity (scale 0–5) of the solution in each tube, using the standard established in Section 5.2. Return each tube to its respective temperature bath immediately after recording.

This experiment addresses the hypothesis that *the temperature of a substrate and an enzyme determines the amount of product that is formed.*

Caution

Wear a heat-resistant glove when handling heated glassware.

11. While you wait for the experiment to run its course, write your prediction of the outcome of the experiment in Table 5-4.
12. Shake periodically (by hand) all tubes over the next 10 minutes.
13. After 10 minutes, remove the test tubes from the water baths. Choose step 14 *or* 15 and proceed.
14. *Color-intensity method:* Record the color intensity (scale 0–5) of each tube's contents in Table 5-4.

TABLE 5-4 Effect of Temperature on Enzyme Activity

Prediction:

Time	Tube 4_A	Tube 4_B	Tube 4_C	Tube 4_D	Tube 4_E	Tube 4_F
			Relative Color Intensity or Absorbance			
0 min.	LIGHT YELLOW	YELLOW	DARK YELLOW	LIGHT YELLOW	CLEAR	
10 min.	YELLOW	DARK YELLOW	PEACH	light peach	clear (1)	

Conclusion:

15. *Spec 20 method:*
    (a) Pipet 6 mL of the contents of each tube into six labeled Spec 20 tubes.
    (b) Clean off each tube.
    (c) Zero the Spec 20 using the contents of tube 2_C.
    (d) Insert tubes containing the contents of each tube, recording them in Table 5-4.
16. Plot the data from Table 5-4 for the 10-minute reading in Figure 5-5.
17. Over what temperature *range* is catechol oxidase active?

_____

18. What is the *optimum* temperature for activity of this enzyme?

_____

19. What happens to enzyme activity at very high temperatures?

DENATURE _____

_____

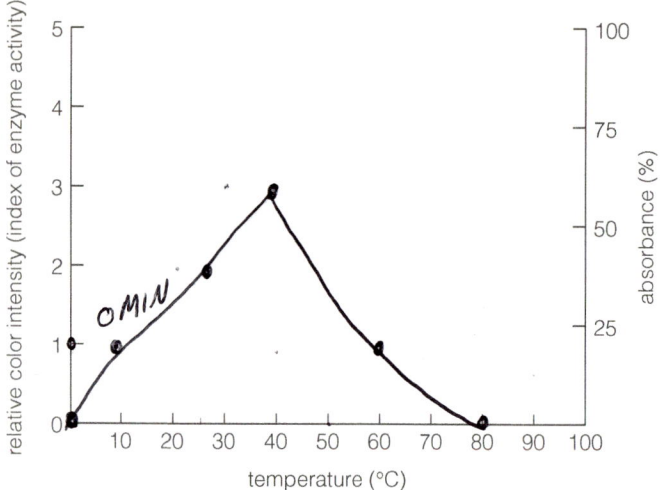

Figure 5-5 Effect of temperature on catechol oxidase activity.

20. Record your conclusion in Table 5-4, either accepting or rejecting the hypothesis.

5.5	**Experiment: Effect of pH on Enzyme Activity** *(25 min.)*

Another factor influencing the rate of enzyme catalysis is the hydrogen-ion concentration (pH) of the solution. Like temperature, pH affects the three-dimensional shape of enzymes, thus regulating their function. Most enzymes operate best when the pH of the solution is near neutrality (pH 7). Others, however, have pH optima in the acidic or basic range, corresponding to the environment in which they normally function.

In this experiment, you will determine the pH range over which the enzyme catechol oxidase is able to catalyze its substrate. You will also determine the optimum pH for the reaction.

MATERIALS

Per student group (4):

- 7 test tubes
- test tube rack
- metric ruler
- china marker
- wash bottle containing 1% catechol
- ice bath with wash bottle of potato extract containing catechol oxidase
- warmed up and zeroed Spec 20, optional; see Appendix 2

Per lab room:

- 40°C waterbath
- phosphate buffer series, pH 2–12 (2, 4, 6, 7, 8, 10, 12)
- vortex mixer (optional)

PROCEDURE

1. With a china marker, label seven test tubes 5_A–5_G. Include your initials for identification.
2. Lay the test tubes against a metric ruler and mark lines indicating 4 cm, 5 cm, and 6 cm *from the bottom* of each tube.
3. Take your test tubes to the location of the phosphate buffer series and fill each tube according to the following directions:

4. Return to your work area and add 1 cm of potato extract containing catechol oxidase to each of the seven tubes (thus bringing the total volume of each to the 5-cm mark). Agitate the tubes by hand.

5. Add 1% catechol to each of the seven tubes, bringing the total volume to the 6-cm mark. Agitate the contents of the tubes, using a vortex mixer if available.

6. In Table 5-5 at time 0, record the relative color intensity of each tube *immediately after adding the 1% catechol.*

7. Place the tubes in the 40°C waterbath.

8. Agitate the tubes periodically over the next 10 minutes.

Tube	Fill to the 4-cm Mark with Buffer of
5_A	pH 2
5_B	pH 4
5_C	pH 6
5_D	pH 7
5_E	pH 8
5_F	pH 10
5_G	pH 12

PURPOSE

This experiment addresses the hypothesis that *the pH of a substrate and an enzyme determines the amount of product that is formed.*

9. While you wait for the experiment to run its course, write your prediction of the outcome in Table 5-5.

10. After 10 minutes, remove the test tubes from the water baths. Choose step 11 or 12 and proceed.

11. *Color-intensity method:* Record the color intensity (scale 0–5) of each tube's contents in Table 5-5.

TABLE 5-5   Effect of pH on Enzyme Activity

Prediction: ·

Time	Relative Color Intensity or Absorbance						
	Tube 5_A (pH 2)	Tube 5_B (pH 4)	Tube 5_C (pH 6)	Tube 5_D (pH 7)	Tube 5_E (pH 8)	Tube 5_F (pH 10)	Tube 5_G (pH 12)
0 min.	CLEAR	CLEAR	LIGHT yellow	yellow	LIGHT yellow	CLOUDY WHITE	PINKISH
10 min.	CLEAR	CLEAR	yellow	peach/brown	PEACH	LIGHT yellow	LIGHT PEACH

Conclusion: 5D

12. *Spec 20 method:*
   (a) Pipet 6 mL of the contents of each tube into seven labeled Spec 20 tubes.
   (b) Clean off each tube.
   (c) Zero the Spec 20 using the contents of tube 2_C from Section 5.2.
   (d) Insert each tube, determine the absorbance, and record it in Table 5-5.

13. Plot the data from Table 5-5 for your 10-minute reading in Figure 5-6.

14. Over what pH *range* does catechol oxidase catalyze catechol to benzoquinone?

   pH 6 - 12

15. What is the *optimum* pH for catechol oxidase activity?

   pH 7

16. Record your conclusion in Table 5-5, either accepting or rejecting the hypothesis.

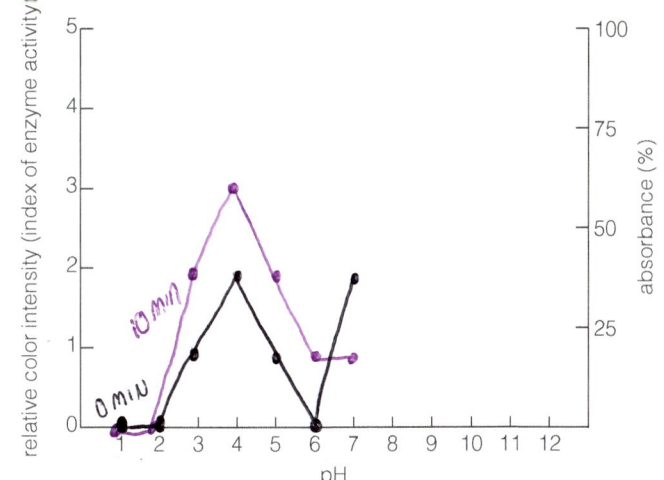

Figure 5-6   Effect of pH on catechol oxidase activity.

Conc.
enzyme - neutral - best state for reaction
more basic/acidic,
   far - dark   close - light

Some enzymatic reactions occur only when the proper *cofactors* are present. **Cofactors** are nonprotein organic molecules and metal ions that are part of the structure of the active site, making the formation of the enzyme–substrate complex possible.

In this experiment, you will use phenylthiourea (PTU), which binds strongly to copper, to remove copper ions. Thus, you'll be able to determine whether copper is a necessary cofactor necessary for producing benzoquinone from catechol.

MATERIALS

Per student group (4):

- 2 test tubes
- test tube rack
- metric ruler
- china marker
- ice bath with wash bottle of potato extract containing catechol oxidase
- wash bottle containing 1% catechol solution
- bottle of dH$_2$O
- china marker
- scoopula (small spoon)

- phenylthiourea crystals in small screw-cap bottle
- warmed up and zeroed Spec 20, optional; see Appendix 2

Per lab room:

- 40°C waterbath
- vortex mixer (optional)
- bottle of 95% ethanol (at each sink)
- tissues (at each sink)

PROCEDURE

1. With a china marker, label two test tubes 6$_A$ and 6$_B$. Include your initials for identification.
2. Lay the test tube against a metric ruler and mark lines indicating 1 cm and 2 cm *from the bottom* of each tube.
3. Add potato extract containing catechol oxidase to the 1-cm mark of each test tube.
4. Using a scoopula, add five crystals of phenylthiourea (PTU) to tube 6$_A$. Do not add anything to tube 6$_B$.

> **Caution**
>
> *PTU is poisonous.*

5. Agitate the contents of both test tubes frequently by hand during the next 5 minutes.
6. Add 1% catechol to the 2-cm mark of both test tubes and agitate the contents of the tubes, using a vortex mixer if available.
7. *Color-intensity method:* In Table 5-5 at time 0, record the relative color intensities (scale of 0–5).

TABLE 5-6 Is Copper a Cofactor for Catechol Oxidase?

Prediction:

	Relative Color Intensity or Absorbance	
Time	Tube 6$_A$: with PTU	Tube 6$_B$: without PTU
0 min.	~~CLEAR~~ CLEAR	~~CLEAR~~ CLEAR
10 min.	CLEAR	yellow
Conclusion:		

---

8. Place the tubes in a 40°C waterbath. Agitate the tubes several times during the next 10 minutes.

This experiment addresses the hypothesis that *a cofactor is necessary for the action of the enzyme catecol oxidase.*

9. While you wait for the experiment to run its course, write your prediction of the outcome in Table 5-6.
10. After 10 minutes, remove the test tubes from the water baths. Choose step 11 *or* 12 and proceed.
11. *Color-intensity method:* Record the color intensity (scale 0–5) of each tube's contents in Table 5-6.

12. *Spec 20 method:*
    (a) Pipet 6 mL of the contents of each tube into seven separate Spec 20 tubes.
    (b) Clean off each tube.
    (c) Zero the Spec 20 using the contents of tube 2_C from Section 5.2.
    (d) Insert each tube in the Spec 20, determine the absorbance, and record it in Table 5-6.

13. Did benzoquinone form in tube 6_A? In tube 6_B? __In 6B, NOT 6A__

14. From this experiment, what can you conclude about the necessity for copper for catechol oxidase activity?
    __NEED COPPER TO form bENZ. - a cofactor__

15. What substance used in this experiment contained copper?
    __POTATOE EXTRACT - PE__

16. Record your conclusion in Table 5-5, either accepting or rejecting the hypothesis.

Note: After completing all experiments, take your dirty glassware to the sink and wash it following directions on page x. Use 95% ethanol to remove the china marker. Invert the test tubes in the test tube racks so they drain. Tidy up your work area, making certain all materials used in this exercise are there for the next class.

_____ 1. Enzymes are
   (a) biological catalysts
   (b) agents that speed up cellular reactions
   (c) proteins
   (d) all of the above

_____ 2. Enzymes function by
   (a) being consumed (used up) in the reaction
   (b) lowering the activation energy of a reaction
   (c) combining with otherwise toxic substances in the cell
   (d) adding heat to the cell to speed up the reaction

_____ 3. The substance that an enzyme combines with is
   (a) another enzyme
   (b) a cofactor
   (c) a coenzyme
   (d) the substrate

_____ 4. Enzyme specificity refers to the
   (a) need for cofactors for some enzymes to function
   (b) fact that enzymes catalyze one particular substrate or a small number of structurally similar substrates
   (c) effect of temperature on enzyme activity
   (d) effect of pH on enzyme activity

_____ 5. For every 10°C rise in temperature, the rate of most chemical reactions will
   (a) double
   (b) triple
   (c) increase by 100 times
   (d) stop

_____ 6. When an enzyme becomes denatured, it
   (a) increases in effectiveness
   (b) loses its requirement for a cofactor
   (c) forms an enzyme–substrate complex
   (d) loses its ability to function

_____ 7. An enzyme may lose its ability to function because of
   (a) excessively high temperatures
   (b) a change in its three-dimensional structure
   (c) a large change in the pH of the environment
   (d) all of the above

_____ 8. pH is a measure of
   (a) an enzyme's effectiveness
   (b) enzyme concentration
   (c) the hydrogen-ion concentration
   (d) none of the above

_____ 9. Catechol oxidase
   (a) is an enzyme found in potatoes
   (b) catalyzes the production of catechol
   (c) has as its substrate benzoquinone
   (d) is a substance that encourages the growth of microorganisms

_____ 10. The relative color intensity used in the experiments of this exercise
   (a) is a consequence of production of benzoquinone
   (b) is an index of enzyme activity
   (c) may differ depending on the pH, temperature, or presence of cofactors, respectively
   (d) is all of the above

EXERCISE 5

Enzymes: Catalysts of Life

5.4 Experiment: Effect of Temperature on Enzyme Activity

1. Eggs can contain bacteria such as *Salmonella*. Considering what you've learned in this exercise, explain how cooking eggs makes them safe to eat.

2. As you demonstrated in this experiment, high temperatures inactivate catechol oxidase. How is it that some bacteria live in the hot springs of Yellowstone Park at temperatures as high as 73°C?

3. Why do you think high fevers alter cellular functions?

4. Some surgical procedures involve lowering a patient's body temperature during periods when blood flow must be restricted. What effect might this have on enzyme-controlled cellular metabolism?

5. At one time, it was believed that individuals who had been submerged under water for longer than several minutes could not be resuscitated. Recently this has been shown to be false, especially if the person was in cold water. Explain why cold-water "drowning" victims might survive prolonged periods under water.

6. Explain what happens to catechol oxidase when the pH is on either side of the optimum.

7. What would you expect the pH optimum to be for an enzyme secreted into your stomach?

Food for Thought

8. Is it necessary for a cell to produce one enzyme molecule for every substrate molecule that needs to be catalyzed? Why or why not?

9. Explain the difference between *substrate* and *active site*.

10. The photo shows slices of two apples. The one on the left sat on the counter for 15 minutes prior to being photographed. The one on the right was sliced immediately prior to the photo being taken.

   a. Explain as thoroughly as possible what you see and why the two slices differ.

(Photo by J. W. Perry.)

   b. If you don't want a cut apple to brown, what can you do to prevent it?

The Digestive Action of Salivary Enzymes on Various Macromolecules

Abstract

Salivary glands secrete enzymes that begin the digestion of some foods within the mouth. Our objective was to determine which macromolecules from various foods the enzymes in saliva would digest. We hypothesized that salivary enzymes would begin digestion of certain carbohydrates and proteins. We used chemical tests to examine the digestive ability of salivary enzymes. Our results indicated that salivary enzymes were capable of digesting starch but were not able to digest glucose or protein. Extensions of these experimental findings are discussed.

Introduction

Enzymes are essential to living organisms. They are catalysts that facilitate and coordinate the metabolic, life-sustaining reactions within cells (Starr, Evers, and Starr 2008). For example, digestive system enzymes help break down food into molecules that are small enough to be absorbed by the body's cells. Enzymes within the cells complete the breakdown of molecules to even smaller units, freeing the energy held within chemical bonds so that it can be stored and later used by the cell (Starr, Evers, and Starr 2008).

To better understand the human digestive system and its enzymatic functions, we undertook a study to investigate the digestive properties of the enzymes contained in saliva. Our research question focused on the function of saliva: What macromolecules does saliva digest? Previous research on mammals has shown that some starches begin to be digested in the mouth (Pederson et al. 2002). To verify this research finding and to extend it by investigating proteins, we designed an experiment to examine the digestion of glucose, starch, and protein by salivary enzymes.

Our hypothesis was that saliva contains a broad spectrum of digestive enzymes that start to digest both simple and complex carbohydrates as well as proteins. This hypothesis led to the predictions that saliva will digest (1) glucose, a simple carbohydrate; (2) starch, a complex carbohydrate; and (3) protein.

Methods

General Procedure

We made separate solutions of glucose, potato starch, and egg albumin (protein). Each solution was contained in a separate test tube. We then collected saliva in test tubes by gently spitting into them until we had a sufficient quantity. For our control groups, we used test tubes with water and the particular macromolecule being tested. For our experimental groups, we used test tubes with water, the particular macromolecule, and saliva. We had one additional control, saliva and water, to verify that saliva and water did not test positive for any of the chemical tests used. We used five replicates of each control and experimental group. To allow for potential digestive action to take place in the experimental groups, we waited 10 minutes before conducting the chemical tests.

Test Procedures

To test for the presence of simple sugars, we used the Benedict's test. We expected that control glucose solutions would change from blue to green or reddish brown when heated with Benedict's reagent, but the experimental solutions with glucose and saliva would remain blue when tested if glucose was digested. We added a dropperful of Benedict's solution to each test tube and heated the test tubes in a boiling water bath for three minutes. Data were recorded in the accompanying table.

To test for the presence of starch, we used the iodine test. We expected that the control starch solutions would change from a yellowish-brown color to a bluish-black color when iodine was added, but the experimental solutions with starch and saliva would remain yellowish-brown if starch was digested. Data also were recorded in the table.

To test for the presence of proteins, we used the Biuret test. We expected the control protein solutions would turn a violet color when the Biuret test reagent was added, but the experimental solutions, with protein and saliva, would remain a light blue if protein were digested. These data were recorded in the table.

TABLE 1 Results of chemical tests for glucose, starch, and protein. NA = test not applicable.

Test Solution	Benedict's Test	Lugol's Test	Biuret Test
Control: water	−	−	−
Control: saliva	−	−	−
Control: glucose + water	+	NA	NA
Glucose + saliva	Expect + if saliva does not digest glucose	NA	NA
Control: Starch (amylose) + water	−	+	NA
Starch (amylose) + saliva	Expect + if digested into simple sugars	Expect—if digested	NA
Control: Egg albumin + water	NA	NA	+
Egg albumin + saliva	NA	NA	Expect—if protein digested

Results

The table contains the results of the chemical tests conducted on all control and experimental groups. The data indicate that saliva digests starch (amylose from potatoes) but is unable to digest glucose or protein. All replicates within a group tested identical.

Discussion

The results of our experimental analysis do not support our hypothesis. Saliva was able to digest potato starch but was unable to digest glucose or protein. Glucose, a small molecule that is highly soluble in water, may begin digestion only after being absorbed into cells (intracellular digestion). Perhaps the glycolytic enzymes are only found within cells and are not excreted for extracellular digestion as are other enzymes of the digestive system. Proteins, on the other hand, are known to be digested by enzymes that are excreted by cells lining the stomach and small intestine (Starr, Evers, and Starr 2008). As the results of this study suggest, proteins may begin digestion only after reaching the stomach, with mastication and no chemical digestion occurring in the mouth.

Future studies that explore the potential effects of temperature, pH, and salinity would add to our knowledge of the digestive action of saliva. Weak bonds, such as the hydrogen bonds that hold enzymes in their functional shapes, may be broken or disrupted by changes in temperature or salinity. Perhaps the old adage "starve a fever but feed a cold" is because the higher body temperatures experienced during a fever disrupt the hydrogen bonds of the body's enzymes, interfering with digestion. Salinity changes may also disrupt the three-dimensional shape of enzymes. Studies that examine the effect of foods high in salt are warranted. Changes in pH may also influence the digestive action of saliva. Beverages such as tea and coffee that are high in acidity may interfere with the digestive ability of saliva and impair digestion in the mouth.

Additional studies could be designed to investigate potential enzyme inhibitors, enzyme activators, and the influence of cofactors on the digestive action of saliva. Many drugs are enzyme inhibitors, as are some herbicides and pesticides. Could pesticide residue on foods reduce the digestive action of saliva? Also, many plants have polyphenolic allelochemicals, such as tannins, that may reduce the digestive action of saliva. A study by Robbins et al. (1987) found differences among herbivores in the ability of their saliva to neutralize tannins. A study to investigate the influence of various polyphenolic allelochemicals would be useful for extending our knowledge of salivary enzyme functioning.

To extend our understanding of enzyme functioning in general, studies could be designed to test the function of over-the-counter digestive enzyme supplements. Various experimental conditions could be created, such as those discussed above for enzymes in saliva, to study the effects of temperature, pH, salinity, enzyme inhibitors and activators, and polyphenolic allelochemicals on enzymatic functioning.

Citations

Pederson, A. M., A. Bardow, S. Beier Jenson, and B. Nauntofte. 2002. Saliva and gastrointestinal functions of taste, mastication, swallowing and digestion. *Oral Diseases* 8(3): 117–29.

Robbins, C. T., S. Mole, A. E. Hagerman, and T. A. Hanley. 1987. Role of tannins in defending plants against ruminants: Reduction in dry matter digestion. *Ecology* 68(6): 1606–15.

Starr, C., C. A. Evers, and L. Starr. 2008. *Biology: Concepts and application,* 7th ed. Belmont, CA: Cengage/Brooks Cole.

Structure and Function of Living Cells

JOURNI   NORMAN [handwritten]

OBJECTIVES

After completing this exercise, you will be able to

1. define *cell, cell theory, plasma membrane, DNA, cytoplasm, prokaryotic, eukaryotic, nucleus, organelle, ribosome, cyanobacteria, cytoplasmic streaming, sol, gel, envelope, mitochondrion, endoplasmic reticulum (rough and smooth), Golgi body, chloroplasts, cell wall, central vacuole;*

2. list the structural features shared by all cells;

3. describe the similarities and differences between prokaryotic and eukaryotic cells;

4. identify the cell parts described in this exercise;

5. state the function(s) for each cell part described in this exercise;

6. describe and identify distinguishing structures between plant and animal cells;

7. identify the structures presented in boldface in the procedure sections.

Introduction

Structurally and functionally, all life has one common feature: all living organisms are composed of **cells**. The development of this concept began with Robert Hooke's seventeenth-century observation that slices of cork were made up of small units. He called these units "cells" because their structure reminded him of the small cubicles that monks lived in. Over the next 100 years, the **cell theory** emerged. This theory has three principles: (1) all organisms are composed of one or more cells; (2) the cell is the basic living unit of organization; and (3) all cells arise from preexisting cells.

Although cells vary in organization, size, and function, all share three structural features: (1) all possess a **plasma membrane** defining the boundary of the living material; (2) all contain a region of **DNA** (deoxyribonucleic acid), which stores genetic information; and (3) all contain **cytoplasm**, which is everything inside the plasma membrane that is not part of the DNA region.

With respect to internal organization, there are two basic types of cells, **prokaryotic** and **eukaryotic**. Study Table 6-1, comparing the more important differences between prokaryotic and eukaryotic cells. The Greek word *karyon* means "kernel," referring to the nucleus. Thus, *prokaryotic* means "before a nucleus," while *eukaryotic* indicates the presence of a "true nucleus." Prokaryotic cells typical of bacteria, cyanobacteria, and archaea are believed to be similar to the first cells, which arose on Earth 3.5 billion years ago. Eukaryotic cells, such as those that comprise the bodies of protists, fungi, plants, and animals, including humans, probably evolved from prokaryotes.

This exercise will familiarize you with the basics of cell structure and function of prokaryotes (prokaryotic cells) and eukaryotes (eukaryotic cells).

[handwritten: P: BACTERIA CYANOBACTERIA ARCHAEA   E: PROTISTS fungi PLANTS Animals inCluDE humans]

6.1   Prokaryotic Cells (*About 20 min.*)

MATERIALS

Per student:
- dissecting needle
- compound microscope
- microscope slide
- coverslip

Per student pair:
- distilled water (dH$_2$O) in dropping bottle

Per student group (table):
- culture of a cyanobacterium (either *Anabaena* or *Oscillatoria*)

Per lab room:
- three bacterium-containing nutrient agar plates (demonstration)
- three demonstration slides of bacteria (coccus, bacillus, spirillum)

TABLE 6-1 Comparison of Prokaryotic and Eukaryotic Cells

Characteristic	Cell Type	
	Prokaryotic	Eukaryotic
Genetic material	Located within cytoplasm, not bounded by a special membrane Consists of a single circular molecule of DNA	Located in **nucleus**, a double membrane-bounded compartment within the cytoplasm Multiple molecules of DNA combined with protein Organized into chromosomes
Cytoplasmic structures	Small ribosomes Photosynthetic membranes arising from the plasma membrane (occur in some types only)	Large ribosomes **Organelles**, multiple kinds of membrane-bounded compartments specialized to perform specific functions
Kingdoms represented	Bacteria Archaea	Protista Fungi Plantae Animalia

© Cengage Learning 2013

PROCEDURE

1. Observe the culture plate with bacteria growing on the surface of a nutrient medium. Can you see the individual cells with your naked eye?

   NO, NEED TO USE A MICROSCOPE

2. Observe the microscopic preparations of bacteria on *demonstration* next to the culture plate. The three slides are made from the three basic shapes of bacteria. Which objective lenses are being used to view the bacteria?

   Objective lens 10x, 40x

   Can you discern any detail within the cytoplasm?

   DIFFERENT SHAPES

   In the space provided in Figure 6-1, sketch what you see through the microscope. Record the magnification you are using in the blank provided in the figure caption.

3. Study Figure 6-2, a three-dimensional drawing of a bacterial cell. Now examine the electron micrograph of the bacterium *Escherichia coli* (Figure 6-3). Locate the **cell wall**, a structure chemically distinct from the wall of plant cells but serving the same primary function to contain and protect the cell's contents.

4. Find the **plasma membrane**, which is lying flat against the internal surface of the cell wall and is difficult to distinguish.

5. Look for two components of the **cytoplasm: ribosomes**, electron-dense particles (they appear black) that give the cytoplasm its granular appearance, and a relatively electron-transparent region (appears light) called the **nucleoid** which contains fine threads of the long DNA molecule.

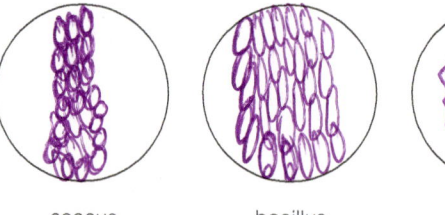

coccus   bacillus   spirillum

Figure 6-1 Drawings of several bacterial cells (10 ×).

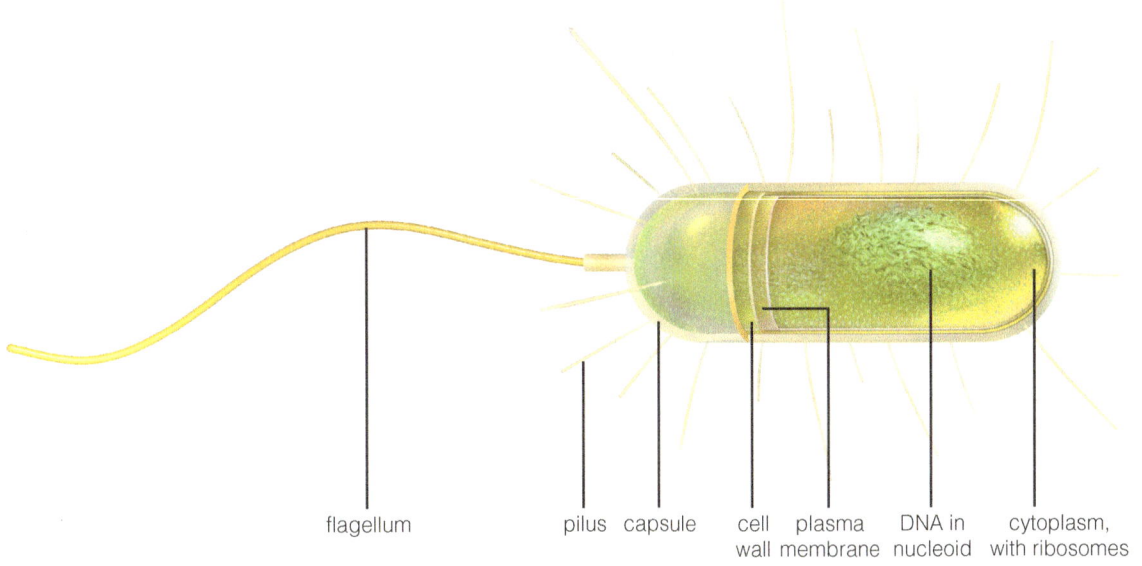

flagellum   pilus capsule   cell   plasma   DNA in   cytoplasm,
                              wall membrane nucleoid with ribosomes

Figure 6-2   Three-dimensional drawing of a bacterial cell as seen with the electron microscope.

Another type of prokaryotic cell is exemplified by cyanobacteria, such as *Oscillatoria* and *Anabaena*. Cyanobacteria (sometimes called blue-green algae) are commonly found in water and damp soils. They obtain their nutrition by converting the sun's energy into chemical energy through photosynthesis.

6. With a dissecting needle, remove a few filaments from the cyanobacterial culture, placing them in a drop of water on a clean microscope slide.

7. Place a coverslip over the material and examine it first with the low-power objective and then using the high-dry objective (or oil-immersion objective, if your microscope is so equipped).

8. In the space provided in Figure 6-4, sketch the cells you see at high power. Estimate the size of a *single* cyanobacterial cell and record the magnification you used to make your drawing.

9. Now examine the electron micrograph of *Anabaena* (Figure 6-5), which identifies the **cell wall, cytoplasm**, and **ribosomes**. These cyanobacteria also possess membranes that function in photosynthesis. Identify the **photosynthetic membranes**, which look like tiny threads within the cytoplasm. Because the electron micrograph of *Anabaena* is of relatively low magnification, the plasma membrane is not obvious, but if you could see it, it would be found just under the cell wall.

10. Look at the captions for Figures 6-3 and 6-5. Judging by the magnification of each electron micrograph, which cell is larger, the bacterium *E. coli* or the cyanobacterium *Anabaena*?

CYANOBACTERIUM ANABAENA

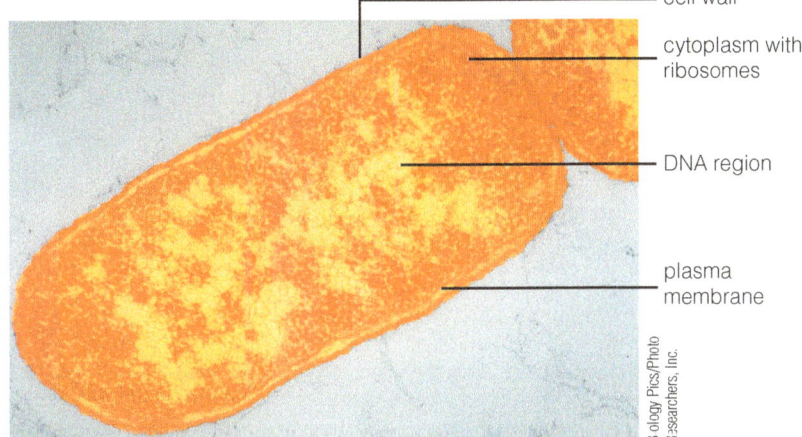

cell wall

cytoplasm with ribosomes

DNA region

plasma membrane

Biology Pics/Photo Researchers, Inc.

Figure 6-3   Electron micrograph of the common intestinal bacterium *Escherichia coli* (28,300×; color added).

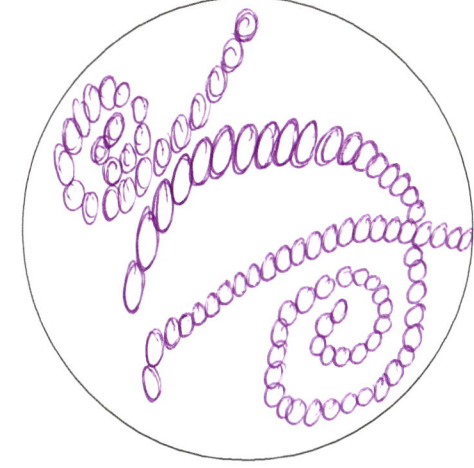

Figure 6-4   Drawing of several cells of a prokaryotic cyanobacterium (_____×).

MATERIALS

Per student:

- textbook
- toothpick
- microscope slide
- coverslip
- culture of *Physarum polycephalum*
- compound microscope
- forceps
- dissecting needle

Per student pair:

- methylene blue in dropping bottle
- distilled water (dH$_2$O) in dropping bottle

Per student group (table):

- *Elodea* in water-containing culture dish
- onion bulb
- tissue paper
- container of ice *or* refrigerator
- Celsius thermometer
- timer or watch with second hand

Per lab room:

- model of animal cell
- model of plant cell

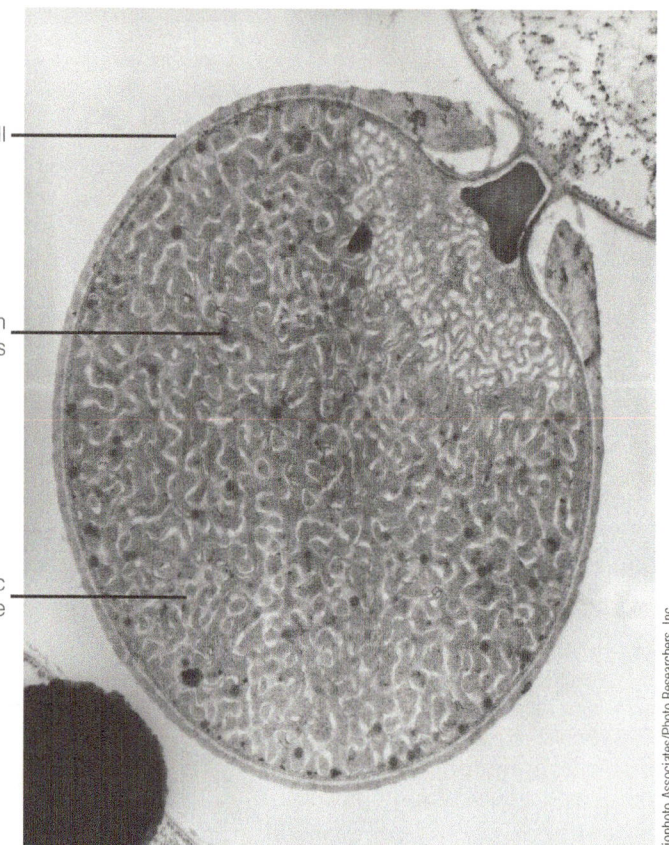

cell wall

cytoplasm with ribosomes

photosynthetic membrane

Biophoto Associates/Photo Researchers, Inc.

Figure 6-5   Electron micrograph of *Anabaena* (11,600×).

A. Cells of a Slime Mold, the Protist, *Physarum polycephalum*

Physarum polycephalum, a member of the Kingdom Protista, is a unicellular organism. As such, it contains all the metabolic machinery for independent existence. Furthermore, its single cell can grow to relatively gigantic size, allowing easy observation of its cytoplasm and cellular organelles.

Physarum absorbs its organic food substances by engulfing particles and digesting them within organelles called *food vacuoles*. Your culture will be growing across the surface of agar gel in a petri dish. The organism grows branches out across the substrate and then comes together on the surface of a food item. The "food" provided to your culture is bacterial cells naturally occurring on the surfaces of oatmeal flakes.

PROCEDURE

1. Place a plain microscope slide on the stage of your compound microscope. This will serve as a platform on which you can place a culture dish.
2. Obtain a petri dish culture of *Physarum*, remove the lid, and place it on the platform. Observe initially with the low-power objective and then with the medium-power objective. If you choose to view the culture with the high-dry objective, place a coverslip over part of the organism before rotating the objective into place. (This prevents the agar from getting on the lens.)

Physarum is *multinucleate*, meaning that more than one nucleus occurs within the cytoplasm. Unfortunately, the nuclei are tiny; you will not be able to distinguish them from other organelles in the cytoplasm.

3. Locate the **plasma membrane**, which is the outer boundary of the cytoplasm. Again, the resolving power of your microscope is not sufficient to allow you to actually view the membrane.
4. Watch the cytoplasm of the organism move. This intracellular motion is called **cytoplasmic streaming**. Contractile proteins called *microfilaments* are believed to be responsible for cytoplasmic streaming. The cytoplasmic motion carries nutrients, proteins, organelles, and other cytoplasmic components throughout the cell.

5. Note that the outer portion of the cytoplasm appears solid; this is the **gel** state of the cytoplasm. Notice that the granules closer to the interior are in motion within a fluid; this interior portion of the cytoplasm is in the **sol** state. Movement of the organism occurs as the sol-state cytoplasm at the advancing tip pushes against the plasma membrane, causing the region to swell outward. The sol-state cytoplasm flows into the region, converting to the gel state along the margins.

6. In Figure 6-6, sketch a portion of *Physarum* that you have been observing and label it.

B. Experiment: Temperature Effects on Cytoplasmic Streaming (About 25 min.)

Temperature affects many cellular and organismal processes. For example, reptiles and insects are ectotherms (animals that gain heat from the environment), unlike humans, whose body heat comes primarily from cellular metabolism. You may have observed that in nature, these animals are relatively sluggish during cold weather. Is the same true for other organisms, such as the slime mold?

This simple experiment addresses the hypothesis that *cold slows cytoplasmic streaming in P. polycephalum.* Before starting this experiment, you may wish to review the discussion in Exercise 1, "The Scientific Method."

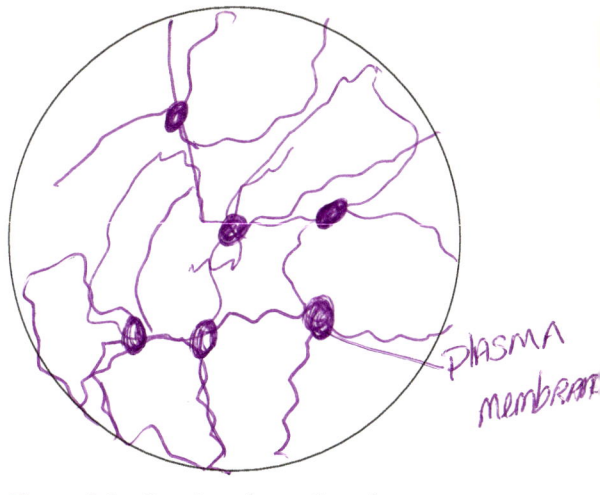

Figure 6-6 Drawing of a portion of *Physarum* (___10___×).

PROCEDURE

1. Place the *Physarum* culture on the stage of your compound microscope as described in Section 6.2.A.
2. Time the duration of cytoplasmic streaming in one direction and then in the other direction. Do this for five cycles of back-and-forth motion. Calculate the average duration of flow in either direction. Record the temperature and your observations in Table 6-2.

TABLE 6-2 Effect of Temperature on Cytoplasmic Streaming

Prediction of observations:

Temperature (°C)	Time/Cycle Number	Observations and Duration of Directional Flow (sec)
	1	
	2	
	3	
	4	
	5	
	Average	
	1	
	2	
	3	
	4	
	5	
	Average	

Conclusions:

© Cengage Learning 2013

3. Remove your culture from the microscope's stage, replace the cover, and place it and the thermometer in a refrigerator or atop ice for 15 minutes.
4. While you are waiting, in Table 6-2, write a prediction for the effect on the duration of cytoplasmic streaming resulting from reducing the temperature of the culture of *P. polycephalum*.
5. After 15 minutes have elapsed, remove the culture from the cold treatment, record the temperature of the experimental treatment, and repeat the observations in step 2.
6. Record your observations and make a conclusion in Table 6-2, accepting or rejecting the hypothesis.

Why was it a good idea to time cytoplasmic streaming for more than a single direction cycle?

TO OBSERVE THE TIME DURATION FOR EACH.

How could this experiment have been improved so that it allows a more reliable conclusion?

USE SAME TEMP FOR ALL TESTS.

A logical question to ask at this time is *why* temperature has the effect you observed. If you perform Exercise 8, "Enzymes: Catalysts of Life," you may be able to make an educated guess (a hypothesis).

C. Human Cheek Cells Observed with the Light Microscope

— EU -
METHYLENE   blue

PROCEDURE

1. Using the broad end of a clean toothpick, gently scrape the inside of your cheek. Stir the scrapings into a drop of distilled water on a clean microscope slide and add a coverslip. Dispose of used toothpicks in the jar containing alcohol.
2. Because the cells are almost transparent, decrease the amount of light entering the objective lens to increase the contrast. (See Exercise 3, page 33.) Find the cells using the low-power objective of your microscope; then switch to the high-dry objective for detailed study.
3. Find the **nucleus,** a centrally located spherical body within the **cytoplasm** of each cell.
4. Now stain your cheek cells with a dilute solution of methylene blue, a dye that stains the nucleus darker than the surrounding cytoplasm and further increases the contrast of the transparent cells. To stain your slide, follow the directions illustrated in Figure 6-7. Without removing the coverslip, add a drop of the stain to one edge of the coverslip. Then draw the stain under the coverslip by touching a piece of tissue paper to the *opposite* side of the coverslip. Search your slide for cheek cells near the central edge of the area colored by the dye.
5. In Figure 6-8, sketch the cheek cells, labeling the **cytoplasm, nucleus,** and the location of the **plasma membrane.** (A light microscope cannot resolve the plasma membrane, but the boundary between the cytoplasm and the external medium indicates its location.) Many of the cells will be folded or wrinkled due to their thin, flexible nature.

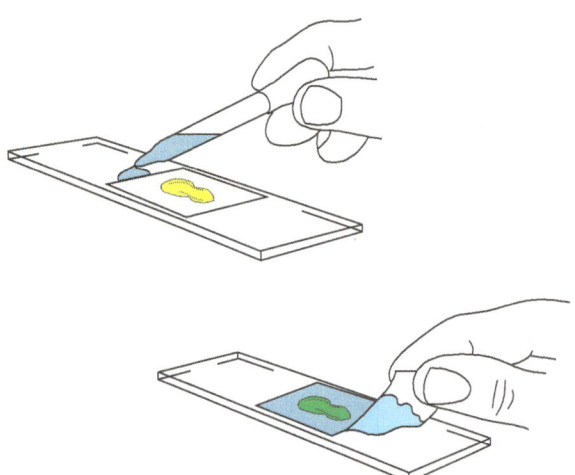

Figure 6-7 Method for staining specimen under coverslip on microscope slide.

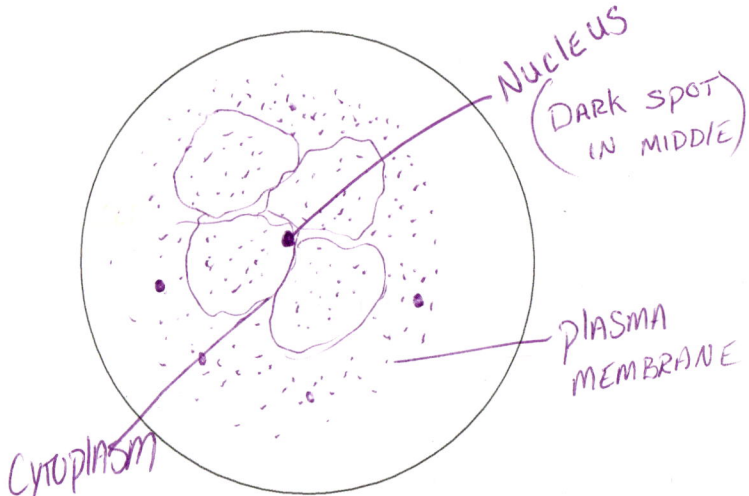

Figure 6-8 Drawing of human cheek cells (_10_ ×).
Labels: cytoplasm, nucleus, plasma membrane

D. Animal Cells as Observed with the Electron Microscope

Studies with the electron microscope have yielded a wealth of information on the structure of eukaryotic cells. Many structures too small to be seen with the light microscope have been identified. These include many **organelles**, structures in the cytoplasm that have been separated ("compartmentalized") by enclosure in membranes. Examples of organelles are the nucleus, mitochondria, endoplasmic reticulum, and Golgi bodies. Although the cells in each of the eukaryotic kingdoms have some unique peculiarities, electron microscopy has revealed that all eukaryotic cells are fundamentally similar.

PROCEDURE

1. Study Figure 6-9, a three-dimensional drawing of an animal cell.
2. With the aid of Figure 6-9, identify the parts on the model of the animal cell that is on *demonstration*.
3. Figure 6-10 is an electron micrograph (EM) of an animal cell (kingdom Animalia). Study the electron micrograph and, with the aid of Figure 6-9 and any electron micrographs in your textbook, label each structure listed.
4. Pay particular attention to the membranes surrounding the nucleus and mitochondria. Note that these two are each bounded by *two* membranes, which are commonly referred to collectively as an **envelope**.
5. Using your textbook as a reference, list the function for the following cellular components:

   (a) plasma membrane _SELECTIVE BARRIER_

   (b) cytoplasm _PROVIDE SUPPORT to INTERNAL STRUCTURE_

   (c) nucleus (the plural is *nuclei*) _HOLD DNA, PROTON + NEUTRONS._

   (d) nuclear envelope _PROTECT NUCLEUS_

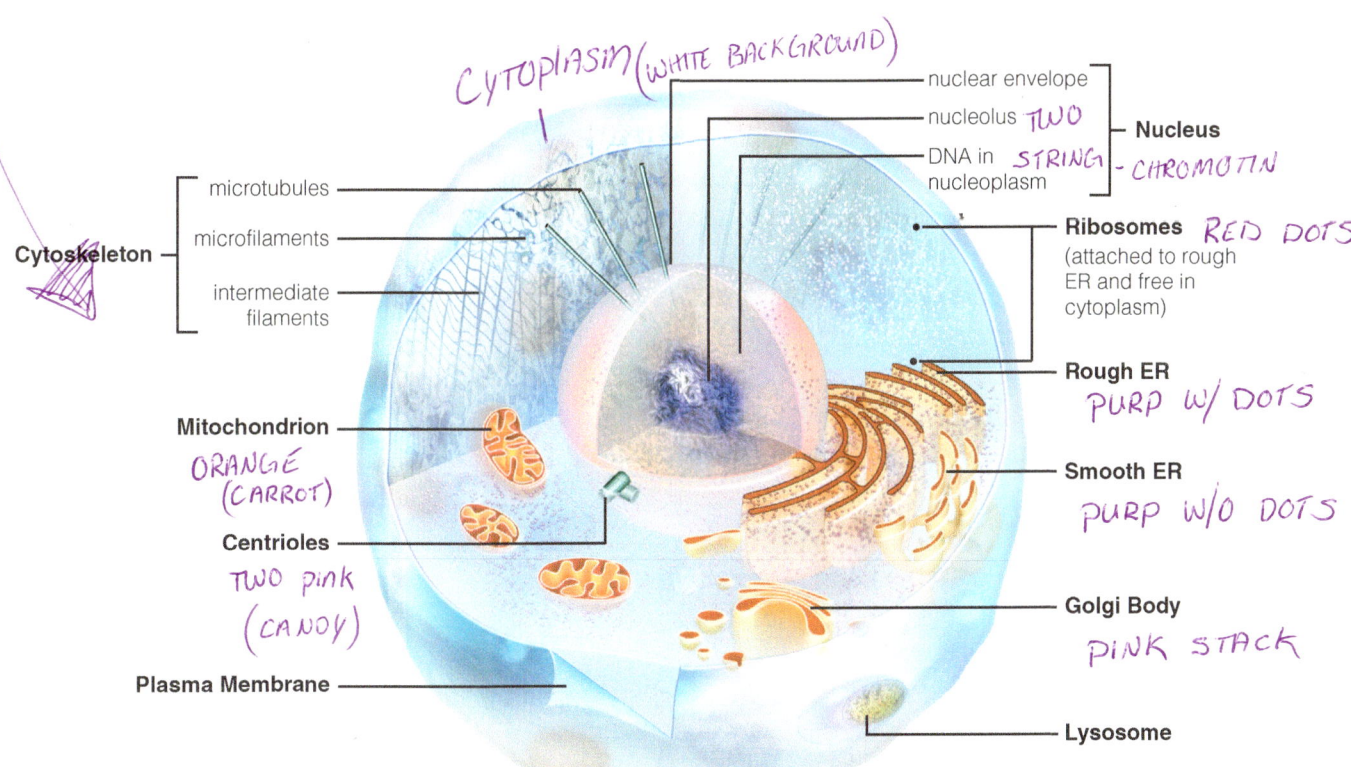

Figure 6-9 Three-dimensional drawing of an animal cell as seen with the electron microscope.

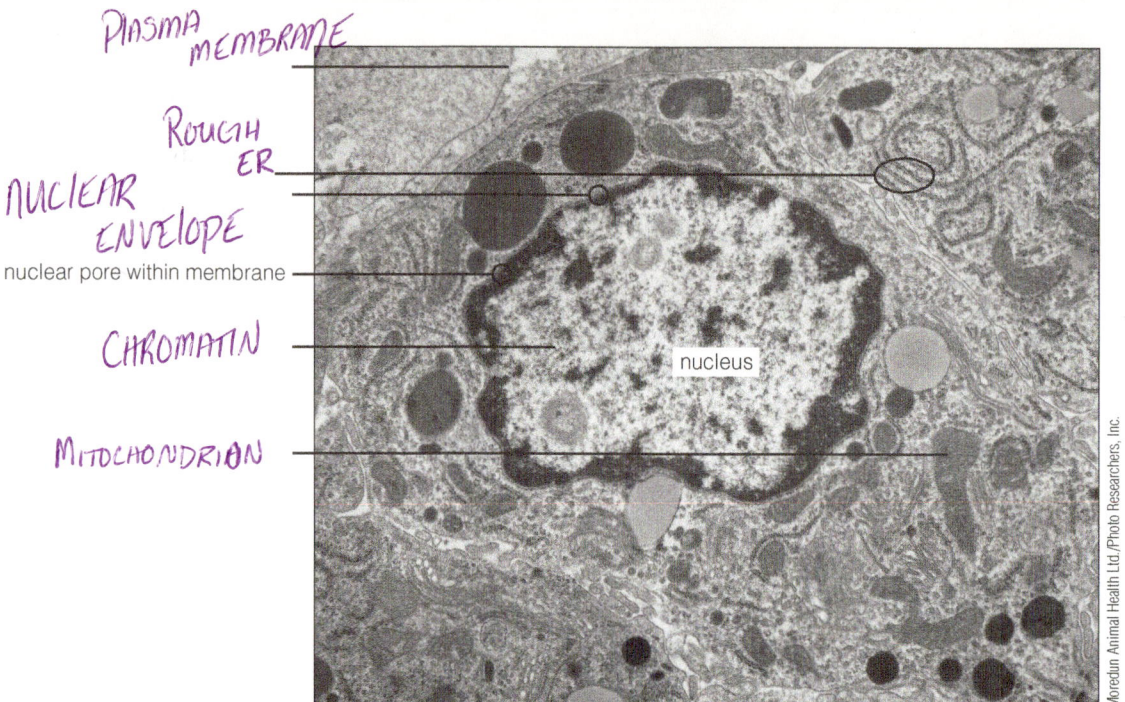

PLASMA MEMBRANE

ROUGH ER

NUCLEAR ENVELOPE

nuclear pore within membrane

CHROMATIN

MITOCHONDRION

nucleus

Moredun Animal Health Ltd./Photo Researchers, Inc.

Figure 6-10 Electron micrograph of an animal cell (1600×).
Labels: plasma membrane, ~~nuclear envelope~~, chromatin, ~~rough ER~~, ~~mitochondrion~~

(e) nuclear pores *ALLOW WATER TRANSPORT*

(f) chromatin *MAKES CHROMOSOMES*

(g) nucleolus (the plural is *nucleoli*) *HOLDS RNA, TRANSCRIBE RNA*

(h) rough endoplasmic reticulum (RER) *PERFORM SYNTHESIS OF PROTEINS*

(i) smooth endoplasmic reticulum (SER) *SYNTHESIS OF CARBS/LIPIDS*

(j) Golgi body *STORE + ROUTE PRODUCTS*

(k) mitochondrion (the plural is *mitochondria*) *POWERHOUSE*

E. Plant Cells Seen with the Light Microscope

E.1. Elodea *leaf cells*

Young leaves at the growing tip of the aquatic pondweed, *Elodea*, are particularly well suited for studying cell structure because these leaves are only a few cell layers thick. Alternatively, your instructor may provide you with similarly thin leaves from a different plant.

PROCEDURE

1. With a forceps, remove a single young leaf, mount it on a slide in a drop of distilled water, and cover with a coverslip.
2. Examine the leaf first with the low-power objective. Then concentrate your study on several cells using the high-dry objective. Refer to Figure 6-11.
3. Observe the abundance of green bodies in the cytoplasm. These are the **chloroplasts**, organelles that function in photosynthesis and that are typical of green plants.
4. Locate the numerous dark lines running parallel to the long axis of the leaf. These are the air-containing *intercellular spaces.*
5. Find the **cell wall**, a structure distinguishing plant from animal cells, visible as a clear area surrounding the cytoplasm.
6. After the cells have warmed a bit, notice the **cytoplasmic streaming** taking place. Movement of the chloroplasts along the cell wall is the most obvious visual evidence of cytoplasmic streaming. Microfilaments (much too small to be seen with your light microscope) are responsible for this intracellular motion.
7. Be aware that you are looking at a three-dimensional object. In the middle portion of the cell is the large, clear **central vacuole**, which can take up from 50% to 90% of the cell interior. Because the vacuole in *Elodea* is transparent, it cannot be seen directly with the light microscope; it simply appears as a large empty space.
8. The chloroplasts occur in the cytoplasm surrounding the vacuole, so they will appear to be in different locations, depending on where you focus in the cell. Focus in the upper or lower surface and observe that the chloroplasts appear to be scattered throughout the cell.
9. Now focus in the center of the cell (by raising or lowering the objective with the fine focus knob), and note that the chloroplasts lie in a thin layer of cytoplasm along the wall. The central vacuole usually pushes the rest of the cytoplasm toward the outer edges of the cell.
10. Locate the **nucleus** within the cytoplasm. It will appear as a clear or slightly amber body that is larger than the chloroplasts. You may need to examine several cells to find a clearly defined nucleus. Often, the clearest clue to the location of a nucleus is a traffic jam of chloroplasts that have clumped against it.
11. Describe the three-dimensional shape of the *Elodea* leaf cell.

    OVAl, OBLONG, CIRCULAR

12. What are the shapes of the chloroplasts and of the nucleus? _CIRCLES - SMAll_

13. Now add a drop of methylene blue stain to make the cell wall more obvious. Add the stain as shown previously in Figure 6-7.
14. Look for the very, very tiny **mitochondria**. (If you have an oil-immersion lens on your microscope, you should use that lens.) They will appear like numerous tiny granules within the cytoplasm.
15. How does the size of the mitochondria compare to that of the chloroplasts?

    MITOCHONDRIOA IS MUCH BIGGER, IN SIZE

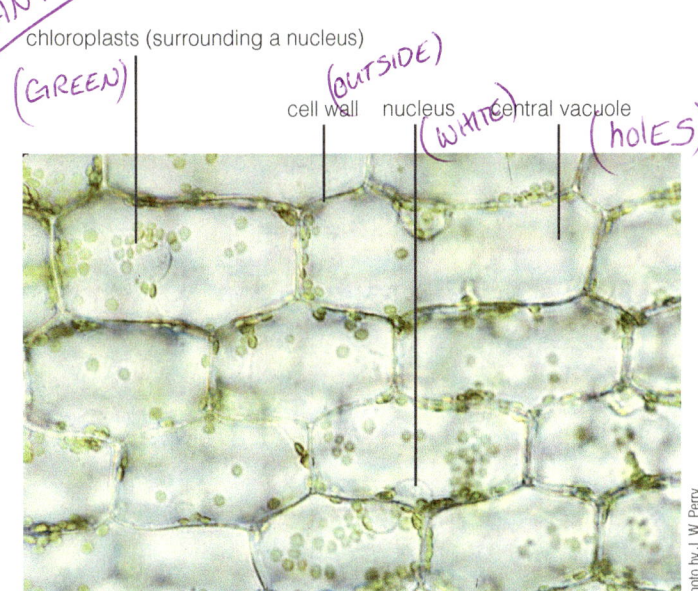

PlANT (GREEN)

chloroplasts (surrounding a nucleus)

(OUTSIDE) cell wall nucleus (WHITE) central vacuole (holES)

Figure 6-11 *Elodea* leaf cells (400×).

Cytoplasm (yellow)

Photo by J. W. Perry

DIff PlANT VS. Animal
    - CENTRIOlES
    - CEll WAll

E.2. Onion scale cells

PROCEDURE

1. Make a wet mount of a leaf of an onion bulb, using the technique described in Figure 6-12. The *inner* face of the leaf is easiest to remove, as shown in Figure 6-12d.
2. Observe your preparation with your microscope, focusing first with the low-power objective. Continue your study, switching to the medium-power and finally the high-dry objective. Refer to Figure 6-13.
3. Identify the **cell wall** and **cytoplasm**.
4. Find the **nucleus**, a prominent sphere within the cytoplasm.
5. Examine the nucleus more carefully at high magnification. Within it, find one or more nucleoli (the singular is *nucleolus*). Nucleoli are rich in a nucleic acid known as RNA (ribonucleic acid), while the nucleus as a whole is largely DNA (deoxyribonucleic acid), the genetic material.
6. You may see numerous *oil droplets* within the cytoplasm, visible in the form of granule-like bodies. These oil droplets are a form of stored food material. You may be surprised to learn that onion "rings" are actually leaves! Which cellular components present in *Elodea* leaf cells are absent in onion leaf cells?

   CHLOROPLASTS

7. If you are using the pigmented tissue from a red onion, you should see a purple pigment located in the vacuole. In this case, the cell wall appears as a bright line.
8. In Figure 6-14, sketch and label several cells from onion bulb leaves.

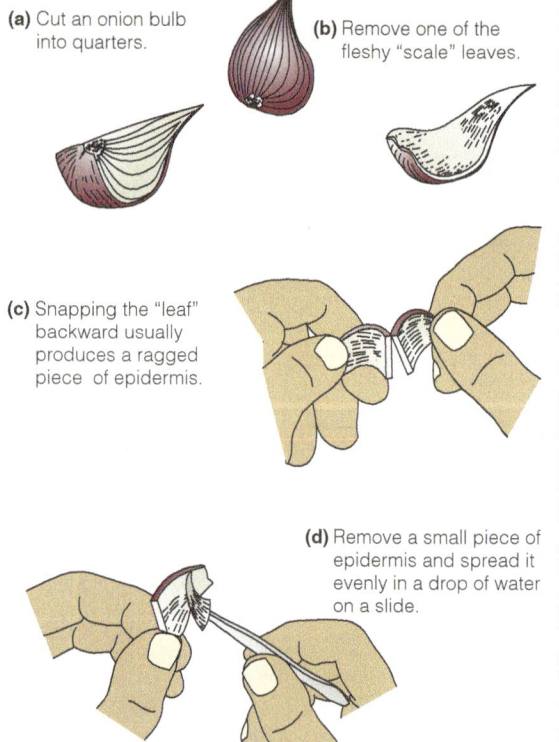

(a) Cut an onion bulb into quarters.

(b) Remove one of the fleshy "scale" leaves.

(c) Snapping the "leaf" backward usually produces a ragged piece of epidermis.

(d) Remove a small piece of epidermis and spread it evenly in a drop of water on a slide.

(e) Gently lower a coverslip to prevent trapping air bubbles. Examine with your microscope. Add more water to the edge of the coverslip with an eye dropper if the slide begins to dry.

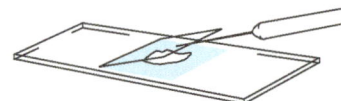

Figure 6-12 Method for obtaining onion bulb leaf cell specimens.

From Peter Abramoff and Robert G. Thomson, Laboratory Outlines in Biology III. Copyright © 1964, 1966, 1972, 1962, 1963 Peter Abramoff and Robert G. Thomson. Copyright © 1982 W. H. Freeman and Company. Used by permission.

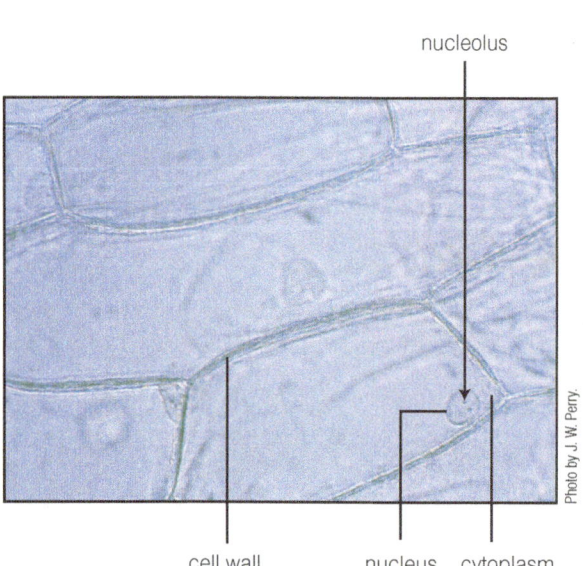

nucleolus

cell wall   nucleus   cytoplasm

Photo by J. W. Perry.

Figure 6-13 Onion bulb leaf cells (67×).

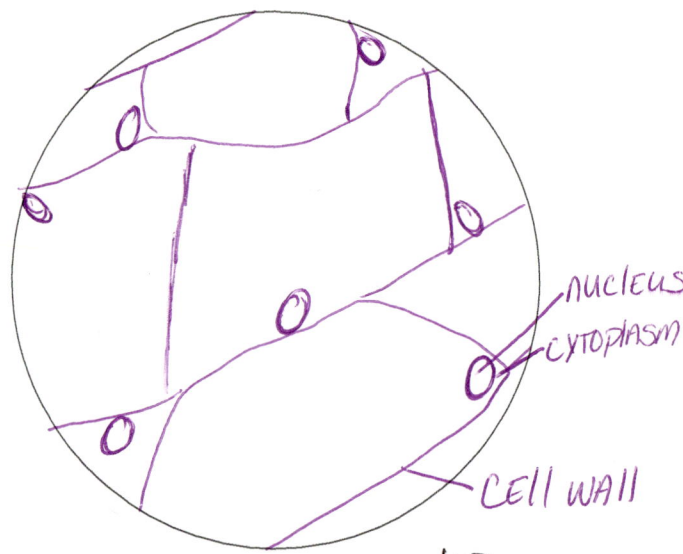

nucleus
cytoplasm
cell wall

Figure 6-14 Drawing of onion leaf cells (40 ×).
Labels: cell wall, cytoplasm, nucleus

F. Plant Cells as Seen with the Electron Microscope

The electron microscope has made obvious some of the unique features of plant cells.

PROCEDURE

1. Study Figure 6-15, a three-dimensional drawing of a typical plant cell.
2. With the aid of Figure 6-15, identify the structures present on the model of a plant cell that is on *demonstration.*
3. Now examine Figure 6-16, a transmission electron micrograph from a corn leaf. Label all of the structures listed. *Caution: Many plant cells do not have a large central vacuole. This is one of them.* Notice that the chloroplast has an envelope, just as do the nucleus and mitochondria.
4. With the help of Figure 6-15 and any transmission electron micrographs and text in your textbook and/or websites, list the function of the following structures:

   (a) cell wall *ACT AS PRESSURE VESSEL, PREVENT OVER EXPANSION*

   (b) chloroplast *TRANSFER CARBS*

   (c) vacuole *ISOLATE subst. (harmful), maintain internal pH*

   (d) vacuolar membrane *PROTECTS VACUOLE, STORES liquid*

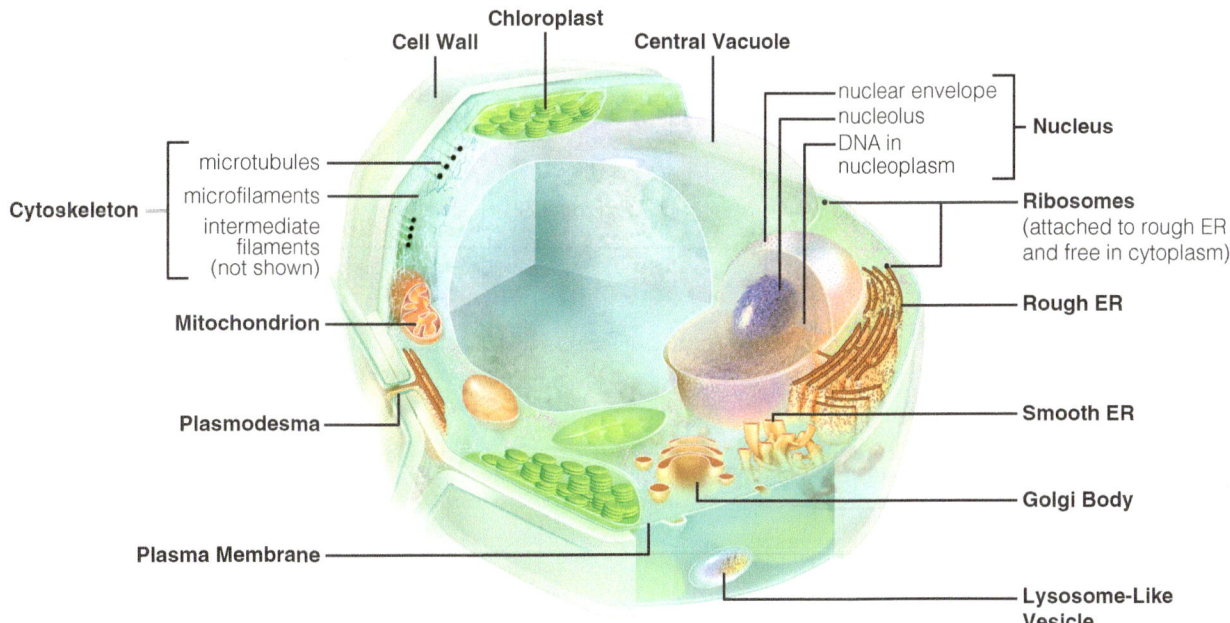

Cell Wall — **Chloroplast** — **Central Vacuole**

nuclear envelope
nucleolus
DNA in nucleoplasm — **Nucleus**

Cytoskeleton
- microtubules
- microfilaments
- intermediate filaments (not shown)

Ribosomes (attached to rough ER and free in cytoplasm)

Mitochondrion

Rough ER

Plasmodesma

Smooth ER

Golgi Body

Plasma Membrane

Lysosome-Like Vesicle

Figure 6-15 Three-dimensional drawing of a plant cell as seen with the electron microscope.

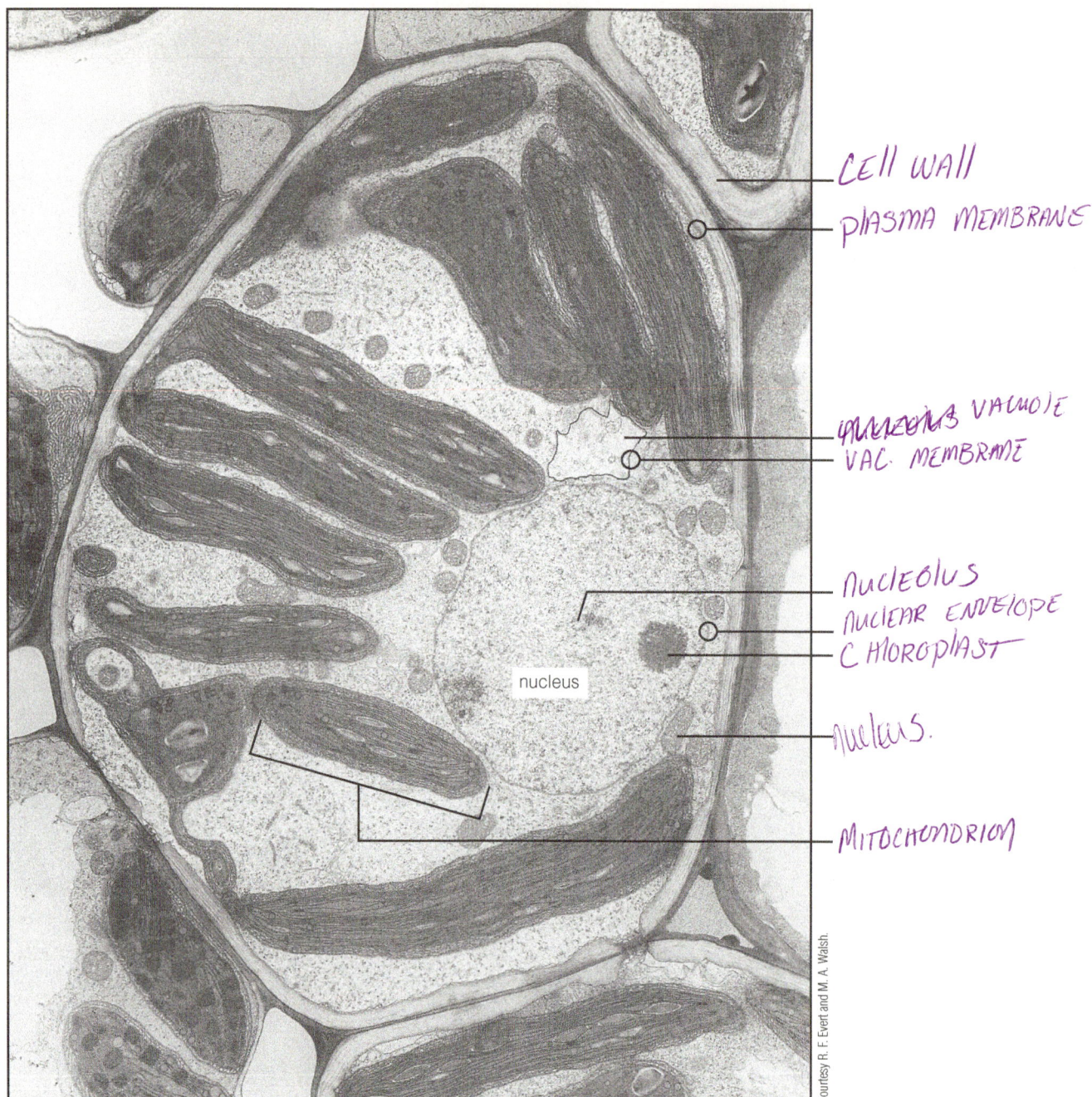

Handwritten labels (purple): CELL WALL, PLASMA MEMBRANE, nucleus VACUOLE, VAC. MEMBRANE, nucleolus, NUCLEAR ENVELOPE, CHLOROPLAST, nucleus, MITOCHONDRION

nucleus

Figure 6-16 Electron micrograph of a corn leaf cell (2700×).
Labels: cell wall, chloroplast, vacuole, vacuolar membrane, plasma membrane, nuclear envelope, chromatin, nucleolus, mitochondrion

Courtesy R. F. Evert and M. A. Walsh.

1. The person who first used the term *cell* was
   (a) Darwin.
   (b) Leeuwenhoek.
   (c) Hooke.
   (d) Watson.

 2. All cells contain
   (a) a nucleus, plasma membrane, and cytoplasm.
   (b) a cell wall, nucleus, and cytoplasm.
   (c) DNA, plasma membrane, and cytoplasm.
   (d) mitochondria, plasma membrane, and cytoplasm.

 3. Prokaryotic cells *lack*
   (a) DNA.
   (b) a true nucleus.
   (c) a cell wall.
   (d) none of the above

 4. The word *eukaryotic* refers specifically to a cell containing
   (a) photosynthetic membranes.
   (b) a true nucleus.
   (c) a cell wall.
   (d) none of the above

 5. A bacterium is an example of a
   (a) prokaryotic cell.
   (b) eukaryotic cell.
   (c) plant cell.
   (d) all of the above

 6. Methylene blue
   (a) is used to kill cells that are moving too quickly to observe.
   (b) detoxifies cells so they are safe to handle.
   (c) is used as food for *Physarum* cells.
   (d) is a biological stain used to increase contrast of transparent cells.

 7. Components typical of plant cells but **not** of animal cells are
   (a) nuclei.
   (b) cell walls.
   (c) mitochondria.
   (d) ribosomes.

 8. A central vacuole
   (a) is found only in plant cells.
   (b) may take up between 50% and 90% of the cell's interior.
   (c) regulates water balance.
   (d) all of the above

 9. The intercellular spaces between plant cells
   (a) contain air.
   (b) are responsible for cytoplasmic streaming.
   (c) produce energy for the cell.
   (d) contain chloroplasts.

 10. An envelope
   (a) surrounds the nucleus.
   (b) surrounds mitochondria.
   (c) consists of two membranes.
   (d) all of the above

EXERCISE **6**

Structure and Function of Living Cells

Post-Lab Questions

6.1 Prokaryotic Cells

1. Did all living cells that you saw in lab contain mitochondria? If not, explain.

2. Below is a high-magnification photomicrograph of an organism similar to one you observed in this exercise. Each rectangular "box" is a single cell. What normally visible organelle is absent from each cell that makes it "prokaryotic?"

Nucleus

(750×)

6.2 Eukaryotic Cells

3. Is it possible for a cell to contain more than one nucleus? Explain.

4. When students are asked to write an essay explaining how to distinguish between an animal cell and a plant cell, they typically answer that plant cells contain chloroplasts and animal cells do not. If you were the professor reading that essay answer, what sort of grade would you assign, and why?

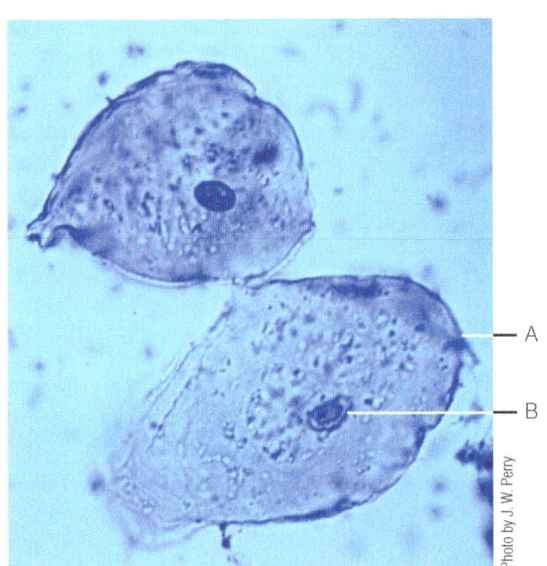

— A

— B

5. Identify the indicated structures.
   A. *PLASMA MEMBRANE*
   B. *NUCLEUS*

6. Look at the photomicrograph to the right, which was taken with a technique that gives a three-dimensional impression. Identify the structures labeled A, B, and C.
   A. <del>Cytoplasm</del> Nucleus
   B. <del>Nucleus</del> Cytoplasm
   C. Cell Wall

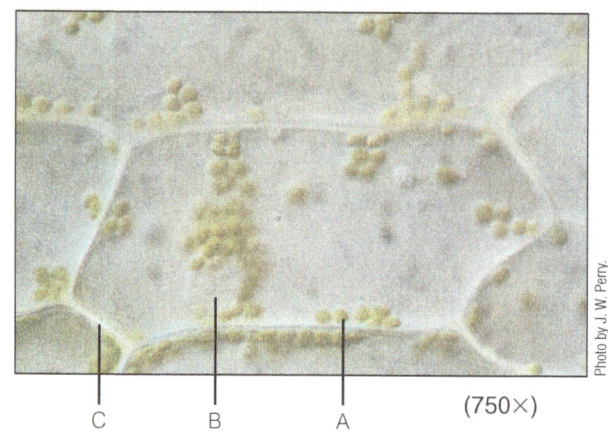

(750×)

7. Describe one function for each structure you identified in (6).
   A. hold important info
   B. Structural support
   C. protect

8. In the electron micrograph below, identify structures labeled A, B, and C.
   A. Nucleus
   B. Mitochondria
   C. Chloroplast

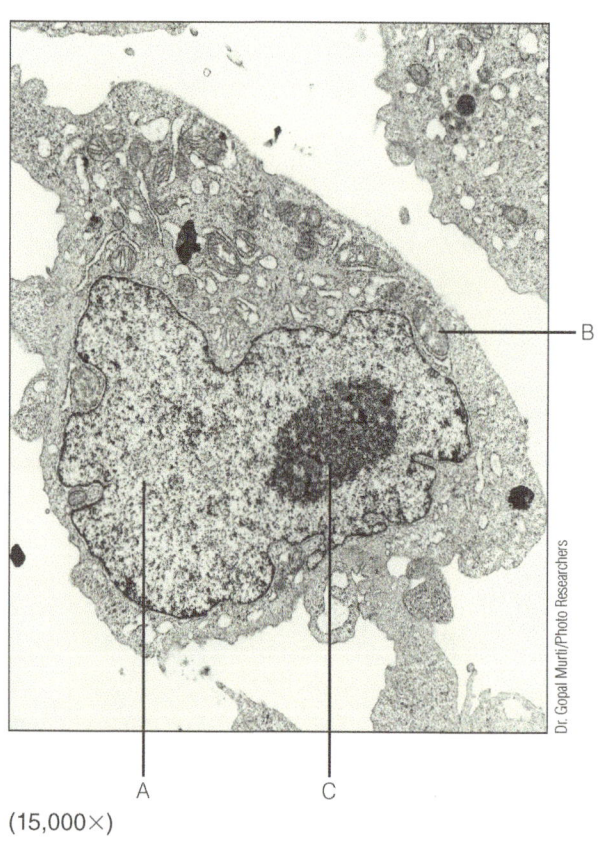

(15,000×)

Food for Thought

9. One botulinum toxin produced by the bacterium *Clostridium botulinum* prevents microfilaments from forming in affected cells. If a *Physarum* cell were exposed to that chemical, would you expect to see cytoplasmic streaming? Explain.

10. Mitochondrial diseases are common, affecting up to 4000 children per year in the United States. Mild forms cause "exercise intolerance." Search two Internet sites for information on "exercise intolerance," and learn about the connections between the role(s) of defective mitochondria and disease symptoms. List the two sites below and briefly summarize the relevant information from each.

http:// _____

_____

_____

_____

_____

_____

http:// _____

_____

_____

_____

_____

_____

EXERCISE 6

Photosynthesis: Capture of Light Energy

OBJECTIVES

After completing this exercise, you will be able to

1. define *photosynthesis, autotroph, heterotroph, chlorophyll, chromatogram, absorption spectrum, carotenoid;*

2. describe the role of carbon dioxide in photosynthesis;

3. determine the effect of light and carbon dioxide on photosynthesis;

4. determine the wavelengths absorbed by pigments;

5. identify the pigments in spinach chloroplast extract;

6. identify the carbohydrate produced in geranium leaves during photosynthesis;

7. identify the structures composing the chloroplast and indicate the function of each structure in photosynthesis.

INTRODUCTION

Photosynthesis is the process by which light energy converts inorganic compounds to organic substances with the subsequent release of elemental oxygen. It may very well be the most important biological event sustaining life. Without it, most living things would starve, and atmospheric oxygen would become depleted to a level incapable of supporting animal life. Ultimately, the source of light energy is the sun, although on a small scale we can substitute artificial light.

Nutritionally, two types of organisms exist in our world, autotrophs and heterotrophs. **Autotrophs** (*auto* means self, *troph* means feeding) synthesize organic molecules (carbohydrates) from inorganic carbon dioxide. The vast majority of autotrophs are the photosynthetic organisms that you're familiar with—plants, as well as some protistans and bacteria. These organisms use light energy to produce carbohydrates. (A few bacteria produce their organic carbon compounds chemosynthetically, that is, using chemical energy.)

By contrast, **heterotrophs** must rely directly or indirectly on autotrophs for their nutritional carbon and metabolic energy. Heterotrophs include animals, fungi, many protistans, and most bacteria.

In both autotrophs and heterotrophs, carbohydrates originally produced by photosynthesis are broken down by *cellular respiration* (Exercise 10), releasing the energy captured from the sun for metabolic needs.

The photosynthetic reaction is often conveniently summarized by the equation:

$$12H_2O + 6CO_2 \xrightarrow{\text{light energy}} 6O_2 + C_6H_{12}O_6 + 6H_2O$$

water carbon dioxide   oxygen glucose water

Although glucose is often produced during photosynthesis, it is usually converted to another transport or storage compound unless it is to be used immediately for carbohydrate metabolism. In plants and many protistans, the most common storage carbohydrate is *starch*, a compound made up of numerous glucose units linked together. Starch is designated by the chemical formula $(C_6H_{12}O_6)_n$, where n indicates a large number. Most plants transport carbohydrate as sucrose.

The following experiments will acquaint you with the principles of photosynthesis.

6.1   Test for Starch *(About 10 min.)*

As indicated by the overall formula of photosynthesis, one end product is a carbohydrate (CH_2O). But a number of different carbohydrates have the empirical formula CH_2O. In this section, you will perform a simple test to visually distinguish between two different carbohydrates and water.

MATERIALS

Per student group (4):

- 1 dropper bottle each of
  iodine (I_2KI) solution
  starch solution

  glucose solution
  dH_2O
- depression (spot) plate

PROCEDURE

1. Place a couple drops of starch, glucose, and distilled water in three different depressions of the spot plate. Now add a drop of iodine solution to each.
2. Record your observations. How can you identify the presence of starch?

   Observations: <u>STARCH PRESENCE WILL CHANGE TO A DARK COLOR.</u>

   _____

   _____

6.2   Experiment: Effects of Light and Carbon Dioxide on Starch Production
(About 30 min.)

In this experiment, you will perform a test to determine the environmental conditions necessary for photosynthesis and starch production. This experiment addresses the hypothesis that *photosynthesis proceeds only in the presence of light and carbon dioxide.*

MATERIALS

Per student group (4):

- two 400-mL beakers
- square of aluminum foil
- hot plate in fume hood
- heat-resistant glove
- petri dish halves
- bottle of iodine solution
- bottle of 95% ethanol (EtOH)
- forceps

Per lab room:

- source of dH_2O
- Fast Plants™, 9–10 days old, grown for 4 days in three different environments:
  - I. normal conditions with both light and carbon dioxide
  - II. in dark, with normal carbon dioxide
  - III. in light, but with carbon dioxide removed

PROCEDURE

1. Carefully observe and record any differences in appearance among the three sets of plants in Table 6-1.
2. Write a prediction regarding starch presence, and thus photosynthesis activity, for each growing condition in Table 6-1.
3. Pigments present in the plants must be removed before a test for starch presence can be performed. Kill the plants and extract the pigments:
   - (a) With a china marker, label the beakers A (for alcohol) and dH_2O.
   - (b) Add about 150 mL distilled water to the dH_2O beaker, set it on the hot plate, and turn on the hot plate to the highest setting. Allow the water to come to a boil.
   - (c) Completely remove 1–2 plants of each treatment from its growing container. Wash off all soil from the roots. Keep plants of each treatment separate and labeled.
   - (d) Alcohol has a much lower boiling point than water, and so takes very little time to come to a boil. When the water is boiling, put about 150 mL of alcohol in the A beaker, set it also on the hot plate and bring to a boil. Keep the alcohol beaker covered with aluminum foil as much as possible throughout the lab to prevent excess evaporation.

   - (e) Place the plants from one treatment in the beaker of boiling water for about 1 minute. This kills the tissue and breaks down internal membranes.
   - (f) Use the long forceps to move the wilted plants from the water into the boiling alcohol. This will extract the photosynthetic pigments from the plant tissues. When the pigments have been extracted, the liquid will appear green, and the plant will appear to be mostly bleached.

> **Caution**
>
> *Ethanol is highly flammable. Use only electric hot plates, never open flame. Also, never let a beaker boil dry. Add more liquid, or remove the beaker from the burner, and place it on a pad of folded paper towels.*

TABLE 6-1	Effect of Light and Carbon Dioxide on Starch Presence		
Fast Plants™ Growing Condition	Appearance	Prediction	Starch Presence and Location
I. Normal conditions with both light and carbon dioxide			
II. In dark, with normal carbon dioxide			
III. In light, but with carbon dioxide removed			

(g) Remove the plants from the alcohol with forceps, and dip momentarily in the boiling water to soften.

4. Test the plants for the presence and localization of starches:
    (a) Place killed, depigmented plants in petri dishes filled with iodine solution.
    (b) Let the plants soak in the iodine solution for a couple minutes, rinse, and float in water in another petri dish in order to observe the pattern of staining.

5. Repeat the pigment extraction an d staining process for plants of the other two treatments, being careful to keep the treatments separate and identified.

6. Remove the A beaker from the hot plate, and turn the heat off.

7. In Figure 6-1, sketch a plant from each experimental treatment, shading in the portions that stained dark. Be careful to note where any dark staining occurs. Record your written observations in Table 6-1.

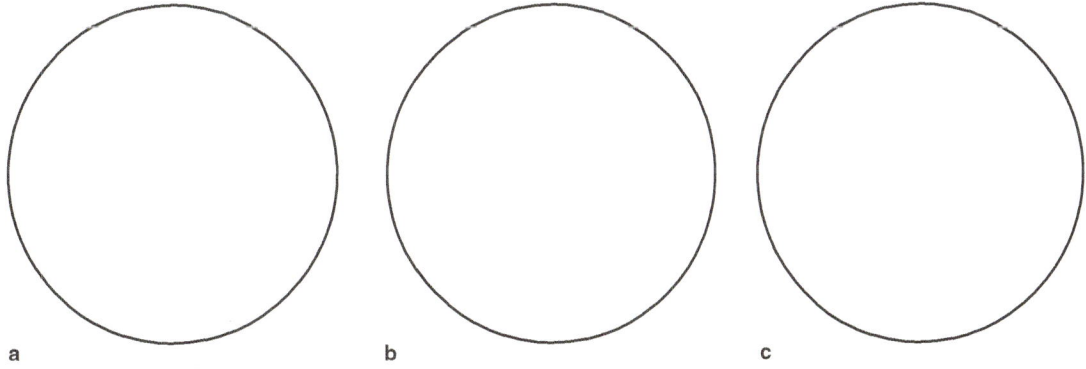

a   b   c

Figure 6-1 Distribution of starch in Fast Plants™. (**a**) Normal light and CO_2; (**b**) dark, normal CO_2; (**c**) light, no CO_2.

8. What does this staining pattern in plants of the three treatments indicate?

_____

_____

_____

9. What conclusion can you draw about the effect of light on the presence of starch?

_____

_____

10. What conclusion can you draw about the effect of carbon dioxide on the presence of starch?

_____

_____

11. Write a conclusion either accepting or rejecting the hypothesis.

_____

_____

LAB REPORT PHOTOSYNTHESIS

6.3	**Experiment: Relationship Between Light and Photosynthetic Products**
	(About 15 min.)

This experiment addresses the hypothesis that *light is necessary for photosynthesis to proceed.*

METHOD

MATERIALS

Per student group (4):

- china marker
- two 400-mL graduated beakers
- hot plate in fume hood
- heat-resistant glove
- bottle of 95% ethanol
- forceps
- 2 petri dishes
- I_2KI solution in foil-wrapped stock bottle

Per lab room:

- source of dH_2O
- light-grown geranium plant or leaves of geranium plant ~~marked "L"~~
- dark-grown geranium plant or leaves of geranium plant ~~marked "D"~~ (kept in dark place)

PROCEDURE

Work in groups of four.

1. Observe the two geranium plants available. One plant has been growing in bright light for several hours; the other has been kept in the dark for a day or more. Both leaves have had an area of the lamina (blade) masked by an opaque design.
2. Write a prediction regarding starch presence, and thus photosynthesis activity, for each growing condition and leaf treatment area in Table 6-2.

TABLE 6-2 Relationship Between Light and Starch Production

Geranium Plant Growing Condition	Prediction		Starch Presence and Location	
	Masked Areas	Unmasked Areas	Masked Areas	Unmasked Areas
Light-grown		YES		Positive
Dark-grown		NO		Negative

3. Select a leaf from one of the two geranium plants. Pigments present in the plants must be removed before a test for starch presence can be performed. Kill the leaf and extract the pigments:
   (a) With a china marker, label the beakers A (for alcohol) and dH_2O.
   (b) Add about 150 mL distilled water to the dH_2O beaker, set it on the hot plate, and turn on the hot plate to the highest setting. Allow the water to come to a boil.
   (c) Remove a treated leaf from each plant. Keep each separate and labeled. Remove the opaque cover from the leaf before proceeding.

(d) Alcohol has a much lower boiling point than water, and so takes very little time to come to a boil. When the water is boiling, put about 150 mL of alcohol in the A beaker, set it also on the hot plate, and bring to a boil. Keep the alcohol beaker covered with aluminum foil as much as possible throughout the lab to prevent excess evaporation.

(e) Place the leaf from one treatment in the beaker of boiling water for about 1 minute. This kills the tissue and breaks down internal membranes.

Caution

Ethanol is highly flammable. Use only electric hot plates, never open flame. Also, never let a beaker boil dry. Add more liquid, or remove the beaker from the burner, and place it on a pad of folded paper towels.

(f) Use the long forceps to move the wilted leaf from the water into the boiling alcohol. This will extract the photosynthetic pigments from the plant tissues. When the pigments have been extracted, the liquid will appear green, and the leaf will appear to be mostly bleached.

(g) Remove the leaf from the alcohol with forceps, dip momentarily in the boiling water to soften.

4. Test the plants for the presence and localization of starches:
   (a) Place killed, depigmented leaf in a petri dish filled with iodine solution.
   (b) Let the leaf soak in the iodine solution for a couple minutes, rinse, and float in water in another petri dish in order to observe the pattern of staining.

5. Show the distribution of the stain in the leaf by shading in and labeling Figure 6-2. In the blank provided in the legend for Figure 6-2, record the *substance* that I_2KI stains.

6. Repeat the pigment extraction and staining process for a leaf of the other treatment. Show the distribution of stain in this leaf by shading in and labeling Figure 6-2.

7. Remove the A beaker from the hot plate, and turn the heat off.

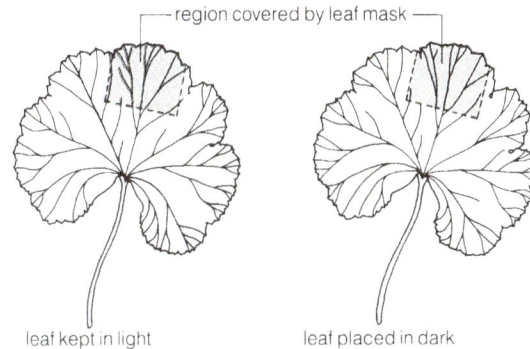

— region covered by leaf mask —

leaf kept in light   leaf placed in dark

Figure 6-2 Distribution of the photosynthetic product ___ *IOx* ___.

What does the blue-black coloration of the leaf indicate? *POSITIVE REACTION FOR STARCH.*

~~Why did the masked area fail to stain?~~

Write a conclusion accepting or rejecting the hypothesis. *Conclusion was correct because STARCH present in the light leaf but NOT THE DARK leaf*

6.4	**Experiment: Necessity of Photosynthetic Pigments for Photosynthesis**
	(About 15 min.)

Coleus plants are widely planted ornamentals that are popular for their striking foliage color patterns. Observe the plants available in the lab and note their wide variety and attractiveness. This experiment addresses the hypothesis that *chlorophyll is necessary for photosynthesis to occur.*

MATERIALS

Per student group (4):

- colored pencils or pens
- two 400-mL beakers
- hot plate in fume hood
- heat-resistant glove
- petri dish halves
- bottle of iodine solution

- bottle of 95% ethanol (EtOH)
- forceps

Per lab room:

- source of dH_2O
- variegated *Coleus* plants

PROCEDURE

1. **Obtain a leaf of variegated *Coleus*.** In the left-hand circle, carefully sketch the leaf, indicating the distribution of each color on the leaf with colored pencils or pens. Green coloration is due to chlorophyll, the major photosynthetic pigment. Pink colors are caused by water-soluble anthocyanin pigments (not involved in photosynthesis), and yellows are formed by carotenoid pigments. Be sure that you look at both surfaces of the leaf in case pigment distribution differs.

   Make a prediction regarding starch presence and photosynthetic activity for each pigmentation area:

   _____

   _____

2. Kill and extract the pigments from the leaf:
   (a) With a china marker, label the beakers A (for alcohol) and dH₂O.
   (b) Add about 150 mL distilled water to the dH₂O beaker, set it on the hot plate, and turn on the hot plate to the highest setting. Allow the water to come to a boil.
   (c) Alcohol has a much lower boiling point than water, and so takes very little time to come to a boil. When the water is boiling, put about 150 mL of alcohol in the A beaker, set it also on the hot plate, and bring to a boil. Keep the alcohol beaker covered with aluminum foil as much as possible throughout the lab to prevent excess evaporation.

   Caution

   Ethanol is highly flammable. Use only electric hot plates, never open flame. Also, never let a beaker boil dry. Add more liquid, or remove the beaker from the burner, and place it on a pad of folded paper towels.

   (d) Place the leaf from one treatment in the beaker of boiling water for about 1 minute. This kills the tissue and breaks down internal membranes.
   (e) Use the long forceps to move the wilted leaf from the water into the boiling alcohol. This will extract the photosynthetic pigments from the plant tissues. When the pigments have been extracted, the liquid will appear green, and the leaf will appear to be mostly bleached.
   (f) Remove the leaf from the alcohol with forceps, and dip momentarily in the boiling water to soften.

3. Test the leaf for the presence and localization of starch:
   (a) Place the killed, depigmented leaf in a petri dish filled with iodine solution.
   (b) Let the leaf soak in the iodine solution for a couple minutes, rinse, and float in water in another petri dish in order to observe the pattern of staining.

4. On the right-hand side of the space below, resketch the leaf, indicating the pattern of staining with iodine.

5. How does the pattern of starch storage relate to the distribution of chlorophyll?

   _____

   _____

6. Write a conclusion either accepting or rejecting the hypothesis.

   _____

   _____

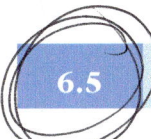
We tend to think of sunlight as being white. However, as you will see in this experiment, white light consists of a continuum of wavelengths. If we see light of just one wavelength, that light will appear colored.

When light hits a pigmented surface, some of the wavelengths are absorbed and others are reflected or transmitted. In this experiment, you will discover *which* wavelengths are absorbed, transmitted, or reflected by *particular* pigments, among them the photosynthetic pigment **chlorophyll.**

MATERIALS

Per lab room:

- several spectroscope setups (Figure 6-3a)
- sets of colored pencils (violet, blue, green, yellow, orange, red)
- colored filters (blue, green, red)
- small test tube containing pigment extract

Per student:

- hand-held spectroscope (optional; Figure 6-3b)

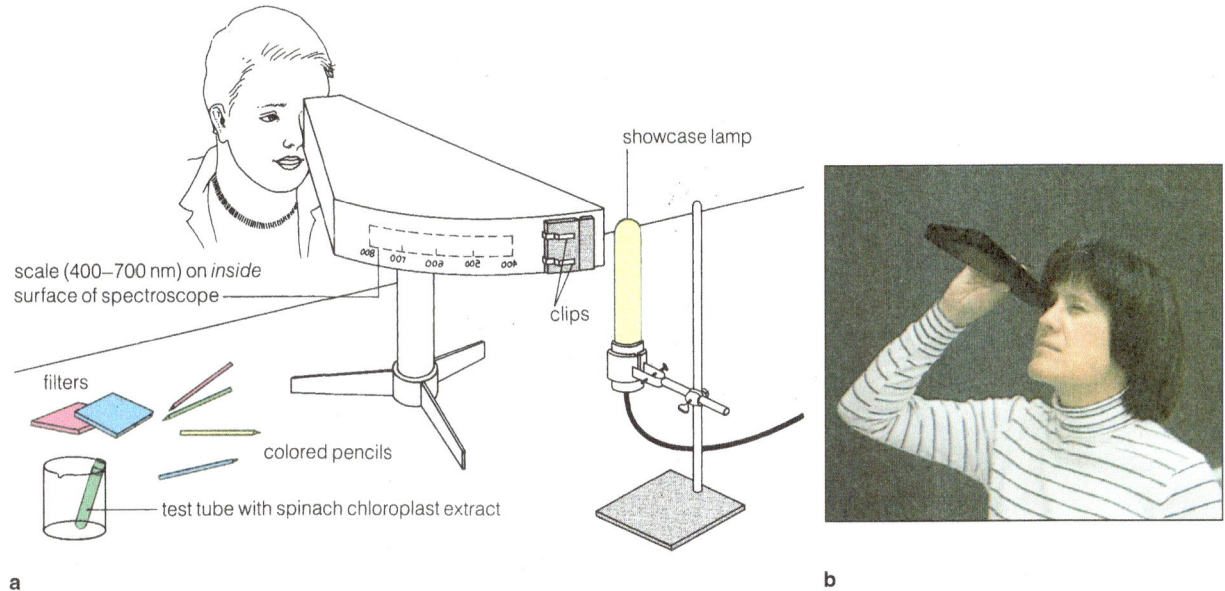

Figure 6-3 Use of a spectroscope. (**a**) Table-mounted. (After Abramoff and Thomson, 1982.) (**b**) Hand-held. (Photo by J. W. Perry.)

PROCEDURE

Work alone.

One way to separate light into its component parts is to view the light through a spectroscope. The spectroscope contains a prism that causes a spectrum of colors to form. A nanometer scale is imposed on the spectrum to indicate the wavelength of each component of white light.

1. Observe the spectrum of white light given off by an incandescent bulb through the spectroscope. With the colored pencils provided, record the positions of the colors violet, blue, green, yellow, orange, and red on the scale in Figure 6-4.
2. Observe the spectrum produced by the three colored filters using the spectroscope.

Which color or colors are absorbed when a red filter is placed between the light and the prism?

 All, EXCEPT RED, IS ABSORBED

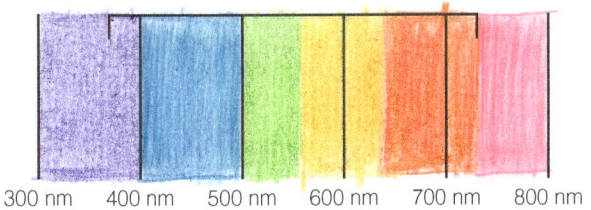

Figure 6-4 Spectrum of white light.

 RED

When a blue filter is used? *All, EXCEPT GREEN & blue, INDIGO*

A green filter? *All, EXCEPT GREEN*

Make a general statement concerning the color of a pigment (filter) and the absorption of light by that pigment.

SHORTER WAVELENGTH = DARKER COLOR, WHILE
longer WAVELENGTH = lighter COLOR.

3. Now obtain a small test tube containing spinach chloroplast pigment extract and place it between the light source and the spectroscope. By adjusting the height of the tube so that the upper portion of the light passes through the pigment extract and the lower portion is white light, you can compare the absorption spectrum of the pigment extract with the spectrum of white light.

An **absorption spectrum** is a spectrum of light waves absorbed by a particular pigment. By contrast, the wavelengths that pass through the pigment extract and are visible in the spectroscope make up the *transmission spectrum* of the pigment.

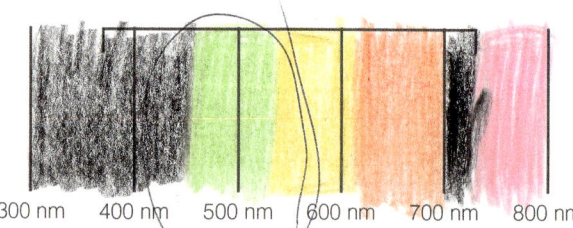

4. Using the colored pencils, record the transmission spectrum of the chloroplast extract on the scale in Figure 6-5.

Figure 6-5 Transmission spectrum of chloroplast extract.

How does the absorption spectrum of the chloroplast extract compare with the absorption spectrum of the green filter?

THERE WAS NO BLUE/INDIGO IN ABSORPTION SPEC.

How might you explain the difference in absorption by the green filter and by the chloroplast pigment extract? (You might want to do Section 6.6 before answering this question.)

ABSORPTION SPEC. ATTRACTS GREEN FILTER.
WHILE CHLORO EXTRACT REFLECT/REJECTS it.

6.6 Separation of Photosynthetic Pigments by Paper Chromatography
(15 min.)

Paper chromatography allows substances to be separated from one another on the basis of their physical characteristics. A chloroplast pigment extract has been prepared for you by soaking spinach leaves in cold acetone and ethanol. Although the extract appears green, other pigments present may be masked by the chlorophyll. In this activity, you will use paper chromatography to separate any pigments present. Separation occurs due to the solubility of the pigment in the chromatography solvent and the affinity of the pigments for absorption to the paper surface. The finished product, showing separated pigments, is called a **chromatogram.**

MATERIALS

Per student:

- chromatography paper, 3 cm × 15 cm sheet
- metric ruler

Per student pair:

- chloroplast pigment extract in foil-wrapped dropping bottle
- chromatography chamber containing solvent
- colored pencils (green, blue-green, yellow, orange)

PROCEDURE

1. Obtain a 3 cm × 15 cm sheet of chromatography paper. *Touch only the edges of the paper,* because oil from your fingers can interfere with development of the chromatogram.
2. Using a ruler, make a *pencil* line (do *not* use ink) about 2 cm from the bottom of the paper.
3. Load the paper by applying a droplet of the chloroplast pigment extract near the center of the pencil line. Allow the pigment spot to dry for about 30 seconds. Five to eight applications of extract on the same spot are necessary to get enough pigment for a good chromatogram. Be certain to allow the pigment to air-dry between applications.

4. Insert your "loaded" chromatography paper, spot-side down, into a chromatography chamber—a bottle containing a solvent consisting of 10% acetone in petroleum ether. The level of the solvent should cover the bottom of the strip but no portion of the pigment spot. Seal the chromatography chamber and allow the solvent to rise on the paper. (Two chromatograms can be inserted in a single bottle, but attempt to keep each separate.)

Caution

Avoid inhaling the solvent vapors. Keep the chamber tightly capped whenever possible.

5. Watch the separation take place over the next 10 minutes. When the solvent is within about 1 cm of the top of the paper, the separation is complete. Remove the strip, close the chromatography chamber, and allow the chromatogram to dry.
6. Using colored pencils, sketch the results in Figure 6-6, showing the relative position of the colors along the paper.
7. Beginning nearest the original pigment spot, identify and label the yellow-green pigment **chlorophyll b.**
8. Moving upward, find the blue-green **chlorophyll a,** two yellow-orange **xanthophylls** in the middle, and an orange **carotene** at the top. Xanthophylls and carotenes belong to the class of pigments called **carotenoids.**
9. You may preserve your chromatogram for future reference by keeping it in a dark place (for example, between the pages of your textbook). Light causes the chromatogram to fade.

Figure 6-6 Chloroplast pigment chromatogram.
Labels: chlorophyll b, chlorophyll a, xanthophylls, carotene

What pigments are contained within the chloroplasts of spinach leaves? GREEN PIGMENTS ARE WITHIN CHLOROPLASTS OF SPINACH LEAVES.

What common "vegetable" is particularly high in carotenes? CARROTS, SWEET POTATOES

(Are you familiar with the medical condition called "carotenosis"? If not, go to the library and look it up in a medical dictionary.)

6.7 **Structure of the Chloroplast** *(15 min.)*

Work alone.

The chloroplast is the organelle concerned with photosynthesis. Study Figure 6-7, an artist's conception of the three-dimensional structure of a chloroplast.

Like the mitochondrion and the nucleus, the chloroplast is surrounded by two membranes. Within the **stroma** (semifluid matrix), identify the **thylakoid disks** stacked into **grana** (a single stack is a **granum**). The chloroplast pigment molecules are located on the surface of the thylakoid disks. Hydrogen ion buildup occurs within the interior of the disks. As these ions are expelled back into the stroma, ATP is formed. Within the stroma, the ATP is used to generate organic compounds. These compounds are converted to carbohydrates, lipids, and amino acids from carbon dioxide, water, and other raw materials.

Now examine Figure 6-8, a high-magnification electron micrograph of a chloroplast. With the aid of Figure 6-7, label the electron micrograph.

If the plant is killed and fixed for electron microscopy after being exposed to strong light, the chloroplasts will contain **starch grains.** Note the large starch grain present in this chloroplast. (Starch grains appear as ellipsoidal white structures in electron micrographs.)

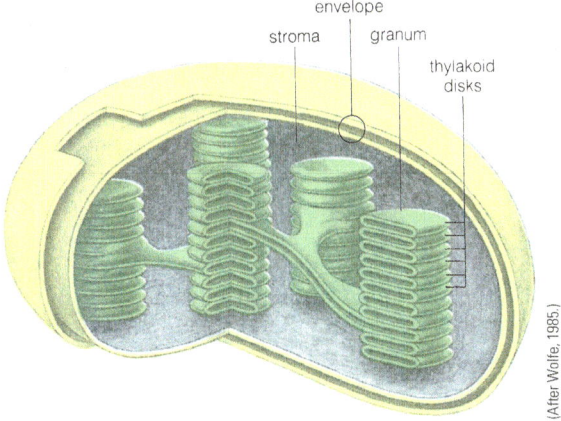

Figure 6-7 The arrangement of membranes and compartments inside a chloroplast.

(After Wolfe, 1985.)

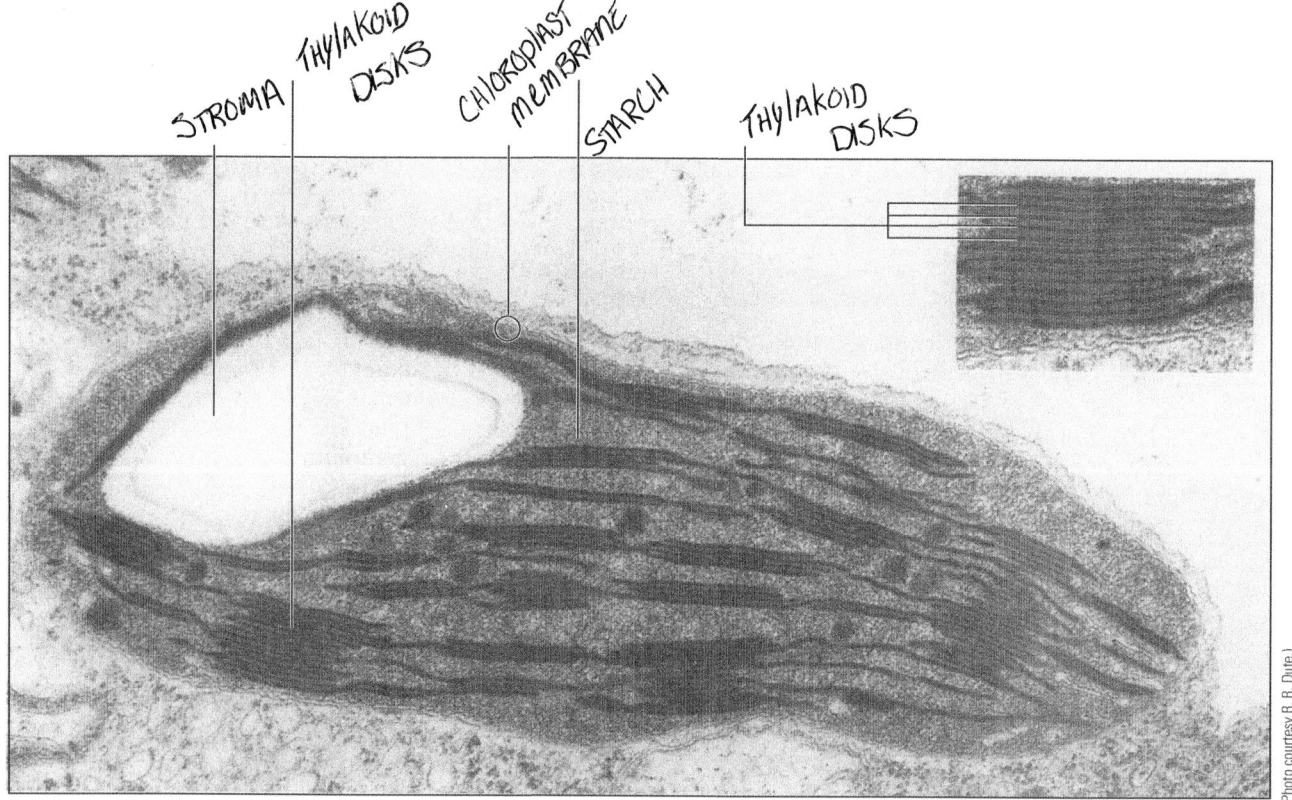

STROMA THYLAKOID DISKS CHLOROPLAST MEMBRANE STARCH THYLAKOID DISKS

(Photo courtesy R. R. Dute.)

Figure 6-8 Electron micrograph of chloroplast (10,000×). Inset: a single granum (20,000×). **Labels:** chloroplast membrane, thylakoid disks, stroma, starch

_C_ 1. The raw materials used for
photosynthesis include
(a) O_2
(b) $C_6H_{12}O_6$
(c) $CO_2 + H_2O$
(d) CH_2O

_A_ 2. A device useful for viewing the spectrum
of light is a
(a) spectroscope
(b) volumeter
(c) chromatogram
(d) chloroplast

_C_ 3. Products and byproducts of
photosynthesis do NOT include
(a) O_2
(b) $C_6H_{12}O_6$
(c) CO_2
(d) H_2O

_C_ 4. A paper chromatogram is useful for
(a) measuring the amount of
photosynthesis
(b) determining the amount of gas
evolved during photosynthesis
(c) separating pigments based on their
physical characteristics
(d) determining the distribution of
chlorophyll in a leaf

_B_ 5. Which of the following pigments would
you find in a geranium leaf?
(a) chlorophyll, xanthophyll,
phycobilins
(b) chlorophyll a, chlorophyll b,
carotenoids
(c) phycocyanin, xanthophyll,
fucoxanthin
(d) carotenoids, chlorophylls,
phycoerythrin

_A_ 6. Which reagent would you use to
determine the distribution of the
carbohydrate stored in leaves?
(a) starch
(b) Benedict's solution
(c) chlorophyll
(d) I_2KI

_B_ 7. An example of a heterotrophic
organism is
(a) a plant
(b) a geranium
(c) a human
(d) none of the above

_A_ 8. Organisms capable of producing their own
food are known as
(a) autotrophs
(b) heterotrophs
(c) omnivores
(d) herbivores

_D_ 9. Grana are
(a) the same as starch grains
(b) the site of ATP production within
chloroplasts
(c) part of the outer chloroplast
membrane
(d) contained within mitochondria and
nuclei

_D_ 10. The ultimate source of energy trapped
during photosynthesis is
(a) CO_2
(b) H_2O
(c) O_2
(d) sunlight

EXERCISE 6

Photosynthesis: Capture of Light Energy

POST-LAB QUESTIONS

6.2 Experiment: Effects of Light and Carbon on Starch Production

1. Is starch stored in the leaves of some plants? Would you expect leaves in a temperate climate plant to be the primary area for long-term starch storage? Why or why not? What part(s) of a plant might be better-suited for long-term starch storage?

6.3 Experiment: Relationship Between Light and Photosynthetic Products

2. Examine the photo at right, which shows the location of starch in two geranium leaves treated in much the same way as you did in Section 9.3. Explain the results you see.

(Photo by J.W. Perry.)

6.4 Experiment: Necessity of Photosynthetic Pigments for Photosynthesis

3. Examine the photo at right of a *Coleus* leaf. Describe an experiment that would allow you to determine whether the deep purple portion of the leaf is photosynthesizing.

(Photo by J.W. Perry.)

6.5 Experiment: Absorption of Light by Chloroplast Extract

4. The photo at right was taken through a spectroscope. What color was the pigment extract used to produce this spectrum? What color(s) did this extract absorb?

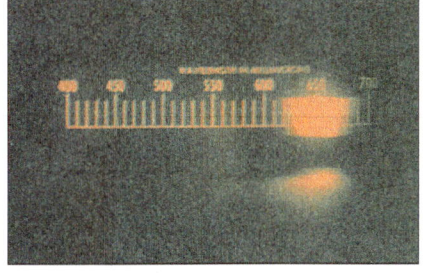

(Photo by J.W. Perry.)

5. Would you illuminate your house plants with a green light bulb? Why or why not?

6.7 Structure of the Chloroplast

6. Examine this electron micrograph of a chloroplast.

   a. Identify the stack of membranes labeled A.

      THYLAKOID DISKS

   b. Identify the region labeled B.

      STROMA

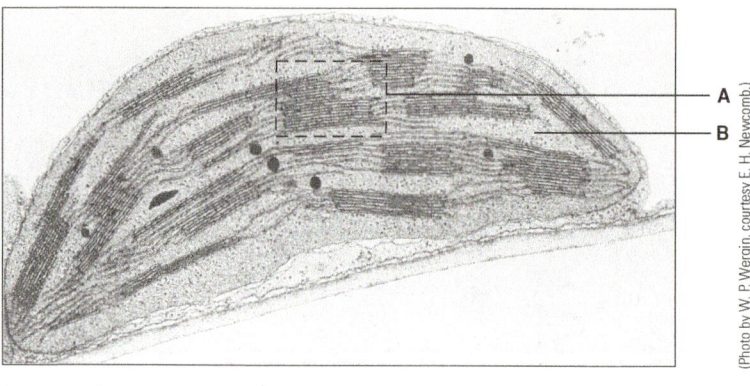

(17,257×).

(Photo by W. P. Wergin, courtesy E. H. Newcomb.)

   c. Would the production of organic compounds during the light-independent reactions occur in region B or on the membranes labeled A?

      B

   d. Would you expect the plant in which this structure was found to have been illuminated with strong light immediately before it was prepared for electron microscopy? Why or why not?

Food for Thought

7. Numerous hypotheses have been proposed for the extinction of the dinosaurs. Recently, evidence has been found of the impact of a large meteor at about the time of this mass extinction. The amount of dust and debris put into the atmosphere upon impact, as well as atmospheric heating, would have been enormous. Using your knowledge of photosynthesis, speculate as to why the dinosaurs subsequently became extinct.

8. Explain the statement: "Without autotrophic organisms, heterotrophic life would cease to exist."

9. Why do you suppose that a chloroplast kept in darkness for some time prior to being fixed for electron microscopy does not contain starch?

10. With the results of the preceding experiments in mind, what might you do to increase the vigor of your house plants?

The Influence of CO_2 Levels on Photosynthesis

Abstract

Understanding the influence of global warming on photosynthetic rates is of paramount importance for predicting ecosystem energy budgets. Global warming involves many environmental factors, not the least of which are rising levels of greenhouse gases such as carbon dioxide (CO_2). The effect of rising CO_2 levels on photosynthesis was examined for one aquatic plant species, *Elodea*. An experimental group given a high concentration of CO_2 was compared to a control group with low levels of CO_2. I hypothesized that the higher CO_2 levels would lead to a lowering of the rate of photosynthesis. This was supported by the experiment. The rate of photosynthesis decreased in the experimental group compared to the control group. Further investigations of the influence of global warming on photosynthesis are suggested and discussed.

Introduction

The photosynthetic efficiency of terrestrial and aquatic species is of central importance for understanding the availability of food and energy within ecosystems. Currently, increases in global warming have led to widespread concern as to its effects on photosynthetic production systems (Chaves and Pereira 1992; Schimel 2006). As such, it is imperative to discover the influence of global warming on photosynthesis. The current study was undertaken to examine the influence of rising CO_2 levels on photosynthesis in an aquatic plant species, *Elodea*. I chose *Elodea* as my study organism because previous research has shown that its photosynthetic rates are reduced under acidic conditions of < pH 6 (Jones, Eaton, and Hardwick 1999). I chose CO_2 levels as my independent variable because this is a leading greenhouse gas contributing to global warming (Lashof and Ahyja 1990), and it may have a detrimental effect on photosynthesis within aquatic ecosystems because it chemically reacts with water, and high levels of CO_2 lead to acidification of aquatic systems (Doney 2006).

As rising CO_2 levels lead to lowering of the pH in aquatic systems, I hypothesized that high CO_2 would lead to a lowering of the pH in an aquatic system and influence the photosynthesis rates of aquatic plant species. I predicted that high levels of CO_2 would lead to a lowering of the photosynthesis rate that would be indicated by a decrease in oxygen production when compared to photosynthesis rates at lower levels of CO_2.

Materials and Methods

Overall Procedure

To measure the rate of photosynthesis, 15 milliliter (ml) centrifuge tubes were filled with water, and a 3 gram (g) sample of *Elodea* was placed inside each. The samples of *Elodea* leaves were matched as closely as possible (size, shape, number, etc). Next, in one-half of the centrifuge tubes, I blew CO_2 through a straw for 5 minutes, using a stopwatch and stopping the timer during pauses where I needed to inhale, covering the top of the centrifuge tube with my thumb to prevent CO_2 from escaping. A rubber stopper fitted with a capillary pipette inserted into a predrilled hole in the stopper was used to plug the centrifuge tubes once I was done. Care was taken to ensure that no air bubbles occurred in the centrifuge tubes once the rubber stopper was fitted. The water level that rose in the capillary pipette after the rubber stopper was fitted was initially marked. It was *assumed* that any water displaced up the capillary pipette column was the result of oxygen being produced by photosynthesis. The capillary tubes were then placed in front of a full-spectrum light source. A period of 15 minutes was allowed to elapse during which time the plants were allowed to acclimate. Then every 10 minutes, the water level was recorded until 30 minutes had elapsed.

Experimental Design

The experimental group for this investigation consisted of three centrifuge tubes with high CO_2 levels, and the control group consisted of three centrifuge tubes with low levels of CO_2. The independent variable was CO_2 levels, and the dependent variable was the displacement of water, which is an indirect measure of the rate of photosynthesis. Two additional controls were used to check the equipment design for the experiment. Two centrifuge tubes only held water and, two centrifuge tubes with just high CO_2 and water were used to ensure that no displacement of water occurred under these conditions.

The average displacement of water was calculated for the experimental and control groups, as well as the standard deviation and standard error of the mean. A t-test was conducted to compare the averages.

Results

The results for the influence of CO_2 levels on photosynthesis in *Elodea* support the hypothesis that high CO_2 levels lead to lower rates of photosynthesis. The *t*-test revealed a significant difference between the control group and the experimental group ($t = 3.0226$, df = 4, $p = 0.02$), therefore the hypothesis was accepted. As depicted in Figure 1 and Table 1, the results were in accordance with what was predicted.

The results for the two additional controls that were used to check the equipment design for the experiment indicated only a slight (0.5 mm) displacement of water on average (mean = 0.5, $SE = 0.2$). This increased the reliability of the data collected.

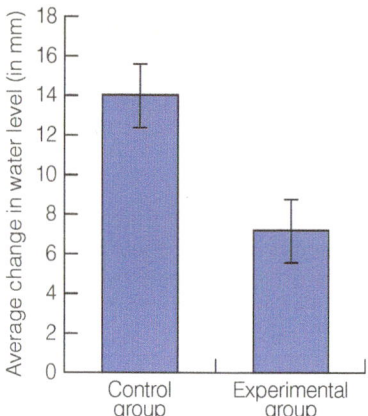

Figure 1 Change in water level (an indirect measure of oxygen production) at different concentrations of CO_2 for *Elodea*. Values are means ± SE.

TABLE 1 Summary Data Table		
Displacement of Water (mm)	Control Group	Experimental Group
Mean	14.00	7.17
Standard Deviation	2.78	2.75
Standard Error	1.61	1.59

Discussion

Photosynthesis was decreased in the experimental group because of the higher levels of CO_2. It was predicted that such higher levels would lead to acidification of the water and lead to reductions in photosynthesis. This was borne out in the experiment. The significance of the results of the current investigation is that even though CO_2 is a reactant molecule for photosynthesis, because of the nature its chemical reactions with water in aquatic systems, too much CO_2 actually decreases photosynthetic rates for pH-sensitive plant species, at least in the short term.

Global warming has serious implications for human societies because of the potential influence on photosynthesis in agricultural crops (Abelson 1992). In addition, rising CO_2 levels have led to increasing acidification of marine environments and the decline of several photosynthetic species such as corals and their symbiotic zooxanthelle (Scott 2006). Thus, it is imperative that we discover the influence of CO_2 levels not only on photosynthesis but also on changes in pH, changes in temperature, changes in sunlight availability, and, for aquatic systems, changes in salinity as well. Future studies that address the influence of these and other factors, as well as examining additional species of photosynthetic organisms, will aid in devising strategies for coping with the effects of global warming.

References

Abelson, P. H. 1992. Agriculture and climate change. *Science* 257: 9.

Chaves, M. M., and J. S. Pereira. 1992. Water stress, CO_2 and climate change. *Journal of Experimental Botany* 43(8): 1131–9.

Doney, S. C. 2006. The dangers of ocean acidification. *Scientific American* March 2006: 58–65.

Jones, J. I., J. W. Eaton, and K. Hardwick. 1999. The effect of changing environmental variables in the surrounding water on the physiology of *Elodea nuttallii*. *Aquatic Botany* 66(2): 115–29.

Lashof, D. A., and D. R. Ahyja. 1990. Relative contributions of greenhouse gas emissions to global warming. *Nature* 344: 529–31.

Schimel, David. 2006. Climate change and crop yields: Beyond Cassandra. *Science* 312: 188–9.

Respiration: Energy Conversion

OBJECTIVES

After completing this exercise, you will be able to

1. define *metabolism, respiration, aerobic respiration, alcoholic fermentation, efficiency, ATP, pH indicator, muscle fiber, ethanol, mitochondrion;*

2. give the overall balanced equations for aerobic respiration and alcoholic fermentation;

3. distinguish among the inputs, products, and efficiency of aerobic respiration and those of fermentation;

4. explain how carbon dioxide input acidifies water and pH indicators can be used to visualize that change;

5. explain how increasing intensity of physical activity affects carbon dioxide production due to aerobic respiration;

6. describe the efficiency of various carbohydrate sources, including corn and sugarcane, for ethanol production through alcoholic fermentation;

7. identify the structures and list the functions of each part of a mitochondrion.

Introduction

All living organisms have a constant energy requirement. Because of this they have mechanisms to gather, store, and use energy. Collectively, these mechanisms are called **metabolism**.

In Exercise 9, we investigated the metabolic pathways by which green plants capture light energy and use it to make carbohydrates such as glucose. Carbohydrates are temporary energy stores. The process by which energy stored in carbohydrates is released to the cell is **respiration**. (Don't confuse this term with the *respiratory cycle* of land vertebrates, during which air is inhaled into lungs and then exhaled.)

Both autotrophs and heterotrophs undergo respiration. Photoautotrophs such as plants use the carbohydrates they have produced by photosynthesis to build new biomass and maintain cellular processes. Heterotrophic organisms may obtain materials for respiration in two ways: by digesting plant biomass or by digesting the tissues of animals that have previously digested plants.

Several different forms of respiration have evolved. The particular respiration pathway used depends on the specific organism and/or environmental conditions. In this exercise, we will investigate two different pathways: (1) **aerobic respiration,** an oxygen-dependent pathway common in most organisms; and (2) **alcoholic fermentation,** an ethanol-producing process occurring in some yeasts (single-celled fungi).

Perhaps the most important aspect to remember about these two processes is that aerobic respiration is by far the most energy-efficient. **Efficiency** refers to the amount of energy captured in the form of ATP relative to the amount available within the bonds of the carbohydrate. **ATP, adenosine triphosphate,** is the so-called universal energy currency of the cell. Energy contained within the bonds of carbohydrates is transferred to ATP during respiration. This stored energy can be released later to power a wide variety of cellular reactions.

For aerobic respiration, the general equation is:

$$\underset{\text{glucose}}{C_6H_{12}O_6} + \underset{\text{oxygen}}{6O_2} \xrightarrow{\text{enzymes}} \underset{\text{carbon dioxide}}{6CO_2} + \underset{\text{water}}{6H_2O} + \underset{\text{chemical energy}}{36ATP^*}$$

If glucose is completely broken down to CO_2 and H_2O, about 686,000 calories of energy are released. Each ATP molecule produced represents about 7500 calories of usable energy. The 36 ATP represent 270,000 calories of energy (36 × 7500 calories). Thus, aerobic respiration is about 39% efficient [(270,000/686,000) × 100%].

By contrast, fermentation yields only 2 ATP for each molecule of glucose entering the process. Thus, fermentation is only about 2% efficient [(2 × 7500/686,000) × 100%]. Obviously, breaking down carbohydrates by aerobic respiration gives a bigger energy payback than through fermentation.

---

*Depending on the tissue, as many as 38 ATP may be produced.

During the process of aerobic respiration, relatively high-energy carbohydrates are broken down in stepwise fashion, ultimately producing the low-energy products of carbon dioxide and water and transferring released energy into ATP. But what is the role of oxygen?

During aerobic respiration, the carbohydrate undergoes a series of oxidation–reduction reactions. Whenever one substance is oxidized (loses electrons), another must be reduced (accept, or gain, those electrons). The final electron acceptor in aerobic respiration is oxygen. Tagging along with the electrons as they pass through the electron transport process are protons (H^+). When the electrons and protons are captured by oxygen, water (H_2O) is formed:

$$2H^+ + 2e^- + \tfrac{1}{2}O_2 \longrightarrow H_2O$$

In the following experiment, we examine aerobic respiration first in germinating seeds and then in humans exercising at different levels.

10.1.A. Experiment: Carbon Dioxide Production (About 25 min. for setup, 1½ hr to complete)

Seeds contain stored food material, usually in the form of some type of carbohydrate. When a seed germinates, the carbohydrate is broken down by aerobic respiration, liberating the energy (ATP) required for each embryo to grow into a seedling.

Two days ago, one set of dry corn seeds was soaked in water to start the germination process. Another set was not soaked and remains dormant. A third set began to germinate but was then boiled to kill them. In this experiment, you will compare carbon dioxide production among germinating corn seeds, ungerminated (dry) corn seeds, and germinating boiled seeds.

This experiment investigates the hypothesis that *germinating seeds produce carbon dioxide from aerobic respiration.*

MATERIALS

Per student group (4):

- three respiration bottle apparatuses (Figure 10-1)
- Sharpie or china marker
- phenol red solution
- 600-mL beaker
- dropper bottle of tetrazolium chloride solution
- petri dish
- single-edged razor blade
- plastic gloves (optional)
- forceps
- magnifying lens *or* dissecting microscope

Per lab room:

- germinating corn seeds
- germinated, boiled seeds
- ungerminated (dry) corn seeds

PROCEDURE

Work in groups.

1. Obtain three respiration bottle setups (Figure 10-1). With a marker, label one "Germ" for germinating corn seeds, a second "Ungerm" for ungerminated seeds, and the third "Boiled" for germinated, boiled seeds.
2. From the class supply, obtain and put enough germinating corn seeds into the "Germ"-labeled respiration bottle to fill it approximately halfway. Fill the "Ungerm" bottle half full with ungerminated (dry) corn seeds. Fill the "Boiled" bottle half full with boiled seeds.
3. Fit the rubber stoppers with attached glass tubes into the respiration bottles. Add enough water to each test tube to cover the end of the glass tubing that comes out of the respiration bottle. (This keeps gases from escaping from the respiration bottle.)
4. Insert a rubber stopper into each thistle tube.
5. Set the bottles aside for the next 1¼ hours and do the other experiments in this exercise.
6. Make a prediction about carbon dioxide production in each bottle and record in Table 10-1.

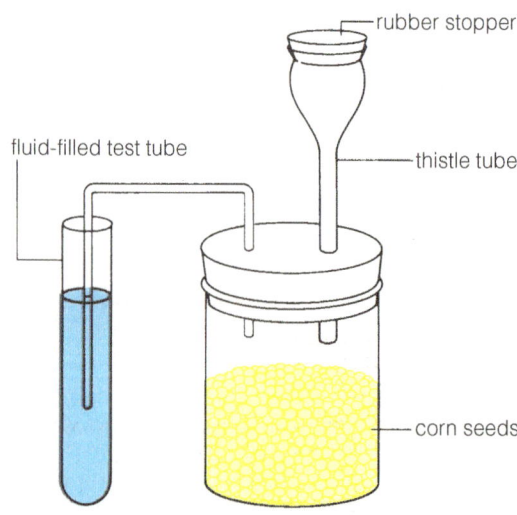

Figure 10-1   Respiration bottle apparatus.

Now start other activities while you allow this one to proceed.

7. After 1¼ hours, pour the water in each test tube into the sink and replace it with an equal volume of dilute phenol red solution. Phenol red solution, which should appear pinkish in the stock bottle, will be used to test for the presence of carbon dioxide (CO_2) within the respiration bottles. If CO_2 is bubbled through water, carbonic acid (H_2CO_3) forms:

$$CO_2 + H_2O \longrightarrow H_2CO_3$$

Phenol red is a **pH indicator**. When the phenol red solution is basic (pH > 8), it is pink; when it is acidic (pH < 6.8), the solution is yellow; and when neutral, it appears orange. The phenol red solution in the stock bottle is _____ BASIC _____ (color); therefore, the stock solution is _____ (acidic/basic/neutral).

8. Put several hundred milliliters of tap water in the beaker.
9. Remove the stopper in the top of the thistle tube and *slowly* pour water from the beaker into each thistle tube. The water will force out gases present in the bottles. If CO_2 is present, the phenol red will become yellow.
10. Record your observations in Table 10-1.

TABLE 10-1	CO_2 Production by Corn Seeds	

Predictions

Germinating corn:

Nongerminating corn:

Boiled corn:

Corn Seeds	Indicator Color (Phenol Red)	Conclusion (CO_2 Present or Absent)
Germinating		
Ungerminated		
Boiled		

Conclusions:

© Cengage Learning 2013

Seeds consist of a protective outer seed coat surrounding an embryo plant and stored food. In a corn grain, the embryo is located just under the seed coat near the pointed end. Most of the interior of the seed is filled with a starchy tissue called endosperm; the endosperm provides a rich source of stored energy for the embryo during germination and its early growth.

Tetrazolium chloride is used by the agricultural seed industry to check the viability of seeds before planting. Tetrazolium is colorless when oxidized but becomes reddish when reduced; the color change happens when a seed's electron transport system of aerobic cellular respiration is working and so indicates a living seed. If the seed is dead, the seeds will remain unstained. We will use tetrazolium chloride to visualize where cellular respiration is proceeding in the germinating corn seeds.

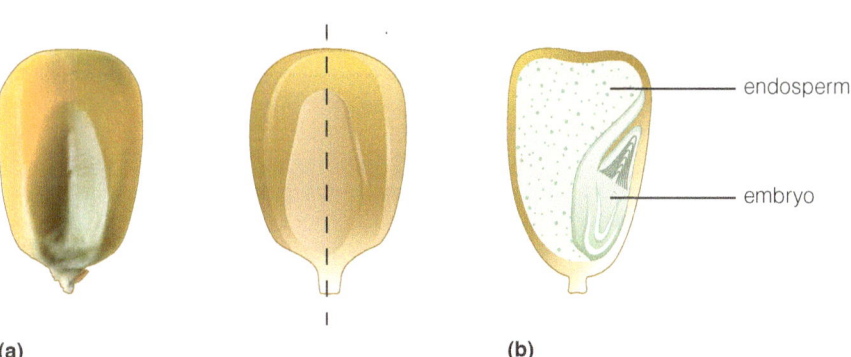

(a) (b)

Figure 10-2 Corn grain (**a**) whole, and (**b**) internal structure.

AGStockUSA/Alamy

11. Remove a grain of germinating corn. Place the kernel flat on a hard surface oriented as in Figure 10-2a and cut in half vertically as indicated by the dashed line.
12. Put several drops of tetrazolium chloride in the petri dish, and use forceps to place the seed halves with the cut side down in the drops. Cover the dishes and allow the seeds to soak for about 15 minutes.
13. Turn the seed halves over and examine them, using a magnifying lens or dissecting microscope. Compare with Figure 10-2b to determine which portions of the seeds have the highest rates of aerobic respiration.

What part(s) of the seed is(are) experiencing the highest rate of respiration?

CAUTION

Tetrazolium chloride is a poison; wear gloves and avoid contact with your skin. Wash immediately and thoroughly if you touch the solution.

10.1.B. Effect of Physical Activity on Cellular Respiration (About 30 min.)

You use large skeletal muscles in your legs, as well as other muscles in your body, when you walk or run. The muscles attach to bones and actively contract, then relax, in order to move the joint (see Exercise 39.) Skeletal muscles are each composed of many long individual muscle cells (fibers). A **muscle fiber** contracts when protein filaments inside slide past each other, shortening the length of the cell. An entire muscle contracts when its component fibers contract in unison.

The sliding action of the filaments is powered by ATP, so muscle fibers have many mitochondria to churn out ATP fuel via cellular respiration. The cells first utilize any glucose circulating in your blood, then break down glycogen, a storage polysaccharide, to produce more glucose. If the physical activity lasts more than about 10 minutes, fats will begin to be broken down to supply energy from that stored in fatty acids.

In this activity you'll see how physical activity affects carbon dioxide production as an indicator of aerobic cellular respiration. You will determine relative carbon dioxide production while one member of your group is at rest, then after a short period of mild exercise, and then again after a short period of moderate activity. What is your prediction regarding a human body's production of carbon dioxide from these three levels of activity? Will more *or* less carbon dioxide be produced as exercise intensity increases? Record in Table 10-2.

We will utilize another pH indicator in this activity. Bromthymol blue (BTB) is blue-colored above pH 7.6, yellow below pH 6.0, and green between pH 6 and 7.6. Recall from Part 10.1.A. that carbon dioxide acidifies water (lowers the pH). We will use the time required for BTB to change from blue to green or yellow as an indication of relative aerobic respiration activity.

MATERIALS

Per student group (4):

- three small beakers
- 10-mL graduated cylinder
- straws
- bromthymol blue (BTB) solution in flask
- large beaker for wastes
- Sharpie or other marker
- stopwatch or clock for timing

PROCEDURE

1. Designate at least one exerciser, a data recorder, an exercise timer, and a setup manager from your team.
2. Label the three small beakers "Trial 1," "Trial 2," and "Trial 3."
3. Measure and pour 10 mL of BTB solution in each beaker.
4. At time 0, the exerciser **slowly** blows through a straw into the bottom of Trial 1 beaker.
5. Repeat the resting control test *twice more*, using the corresponding beakers. Try to be consistent in using approximately the same color endpoint for all timing.
6. Discard used BTB solutions in the waste beaker and replace with fresh BTB.
7. Repeat steps 4 to 6 with the exerciser first walking quickly for one minute around the classroom or in a nearby corridor (mild exercise), and then running in place or up and down a nearby staircase for one minute (moderate activity).
8. Calculate the average of the three trials of each exercise level and record in Table 10-2.

> **CAUTION**
>
> Do not inhale! When the solution changes color to green or yellow, record the elapsed time (in seconds) in Table 10-2.

TABLE 10-2	Effect of Activity Level on Carbon Dioxide Production from Aerobic Respiration				

Exercise Level	Trial 1	Trial 2	Trial 3	Average	Daily Difference
Resting	120			120	
Walking	84			84	
Running	20			20	

Time (seconds) to Color Change of Bromthymol Blue Solution

© Cengage Learning 2013

9. Calculate the ratio between each of the two activity levels and the resting control average (activity level average divided by resting average). Record in Table 10-2.

10. What can you conclude about the effect of activity intensity on aerobic cellular respiration? Record in Table 10-2.

CAUTION

Do not do this activity if you have a medical condition that interferes with your ability to exercise. If you feel weak or dizzy while exercising, stop immediately and sit down.

10.2 Experiment: Alcoholic Fermentation and Ethanol Production
(About 20 min. to set up, 1½ hr to complete)

Alcoholic fermentation is an anaerobic (nonoxygen requiring) process occurring in some yeast and a few bacterial cells. Fermentation provides sufficient energy for those organisms to survive despite relatively low ATP yield; most of the energy originally contained in glucose remains in the ethanol product. The chemical equation for this process is:

$$C_6H_{12}O_6 \longrightarrow 2C_2H_5OH + CO_2 + 2ATP$$

glucose   ethanol   carbon   energy
                dioxide

Alcoholic fermentation by yeast is the basis for the baking, wine-making, and brewing industries. Additionally, fermentation is used industrially to produce ethanol for transportation fuel. Gasoline engines can be modified to operate on blends of gasoline with up to 85% ethanol added, suggesting the possibility of a biofueled transportation future.

The United States and Brazil are the world's leading producers of ethanol. In the United States, corn grain is the major feedstock for the fermentation process. The endosperm of the corn grain (see Figure 10-2) is made mostly of starch. This starch is enzymatically broken down into glucose subunits; yeasts then ferment the glucose into ethanol. In contrast, Brazil produces ethanol from yeast fermentation of sugars extracted directly from sugarcane stalks (sugarcane is a major crop in wet, tropical Brazil.) Other plant materials are also being investigated and used for commercial ethanol production. Especially promising is "cellulosic" ethanol, produced by digesting the cellulose plant cell walls abundant in nonfood materials such as straw and cornstalks into glucose for fermentation. But are all carbohydrate sources equally efficient in yielding bioethanol? This activity will offer insight into some answers to that question.

MATERIALS

Per student group (4):
- Sharpie or other marker
- five 50-mL beakers
- 500-mL flask
- 25-mL graduated cylinder
- 200-mL graduated cylinder
- five fermentation tubes *or* alternative provided by your instructor
- balance and weighing paper
- dry rapid-rise yeast
- NaCl
- spoons *or* metal spatulas
- 15-cm metric ruler

Per lab room:
- glucose
- ground corn grain
- pulverized sugarcane stem
- at least two additional ground carbohydrate materials (All Bran cereal, straw, etc.)
- warm distilled water (about 38°–43°C)
- source of distilled water

PROCEDURE
Work in groups.

1. Prepare the active yeast culture as follows:
   (a) Weigh 0.2 g of NaCl and add to 500-mL flask.
   (b) Weigh 7.0 g of dry rapid-rise yeast and add to flask.
   (c) Measure 200 mL of warm water and add to flask. Swirl to mix.

2. Using a marker, number five 50-mL beakers. Also number the five fermentation tubes.

3. Weigh 2 g of each of the following and place into the corresponding beaker: #2, glucose; #3, ground corn; #4, pulverized sugarcane stem; #5, carbohydrate source of your choice. (Beaker #1 has no carbohydrate added.)

4. With a 25-mL graduated cylinder, measure out and pour 20 mL[†] of yeast solution into each of the five beakers. Swirl to mix.

5. When each is thoroughly mixed, pour the contents into five correspondingly numbered fermentation tubes (Figure 10-3). Cover the opening of the fermentation tube with your thumb and invert each fermentation tube so that the "tail" portion is filled with the solution. This is "time zero" for this experiment.
   What is the purpose of Tube 1 (yeast culture only)? __ETHANOl__
   What is the purpose of Tube 2 (yeast culture plus glucose)? __CO₂__
   What is the independent variable in this experiment? __YEAST__
   What is the dependent variable in this experiment? __GlucoSE / GASES__

6. What gas will be produced by the process of fermentation? __CO₂__ Note that we will be measuring the relative amounts of this gas in each tube as an indirect measure of ethanol production. Write a prediction about relative amounts of gas production in each tube in Table 10-3.

7. At intervals of 10 minutes over the next hour, measure the height of the gas bubble (if any) that forms in the closed end of the tube. Record data in Table 10-3.

8. Graph data from each fermentation tube in Figure 10-4. Be careful to clearly distinguish between treatments.

9. What conclusions can you draw regarding the efficiency of ethanol production with each carbohydrate source in Beakers 3–5 relative to that from glucose in Beaker 2? Record in Table 10-3.

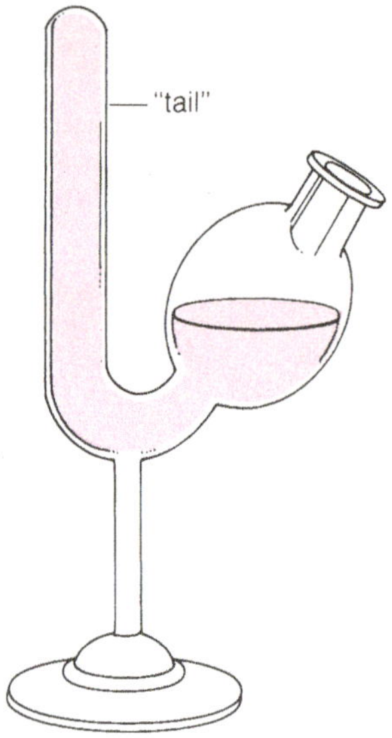

Figure 10-3 Fermentation tube.

[†]The amount of fluid needed to fill the fermentation tube depends on its size. Your instructor may indicate the required volume.

TABLE 10-3	**Production of Gas by Yeasts with Various Carbohydrate Sources**

Predictions

Tube 1: ETHANOl

Tube 2: CARBON DIOXIDE

Tube 3: CARBON DIOXIDE

Tube 4:

Tube 5:

		Distance from Tip of Tube to Fluid Level (mm)					
Tube	Solution	10 min.	20 min.	30 min.	40 min.	50 min.	60 min.
1	Yeast culture						93mm
2	Yeast culture plus glucose						9mm
3	Yeast culture plus ground corn						34mm
4	Yeast culture plus pulverized sugarcane						
5	Yeast culture plus						

Conclusions:

© Cengage Learning 2013

Did your results conform to your predictions? If not, speculate on reasons why this might be so. YES THE RESULTS CONFIRMED MY HYPOTHESIS THAT CO₂ AND ETHANOL WOULD BE PRESENT

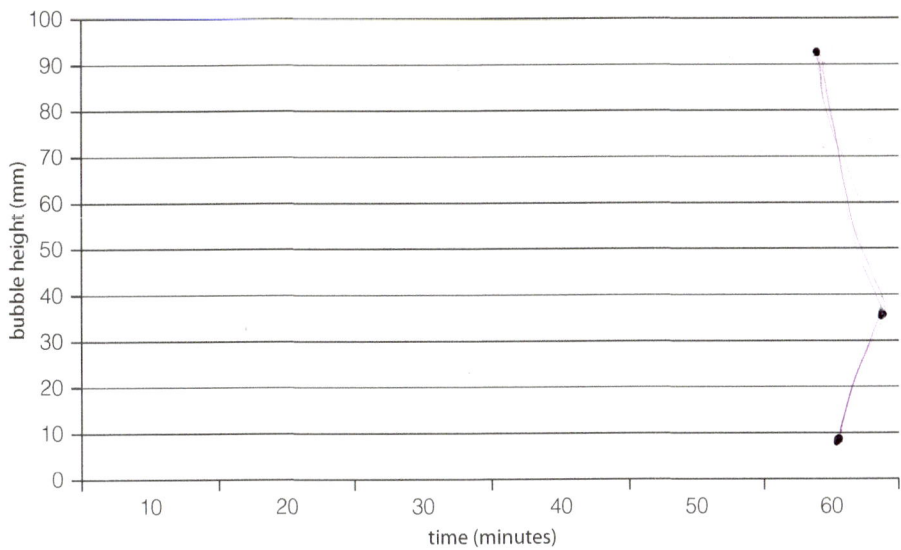

Figure 10-4 Gas production (mm) from fermentation of various carbohydrate sources.

Pool your results with those of other groups in your class. Which carbohydrate source(s) produced the greatest volume of gas, and by extension, the most ethanol? _YEAST_
Which industrial raw material, corn or sugarcane, seems to be the more efficient for ethanol production?
CORN

10.3 Ultrastructure of the Mitochondrion

1. Study Figure 10-5b, which shows the three-dimensional structure of a mitochondrion, the respiratory organelle of all living eukaryotic cells. The mitochondrion has frequently been referred to as the "powerhouse of the cell," because most of the cell's chemical energy (ATP) is produced here.

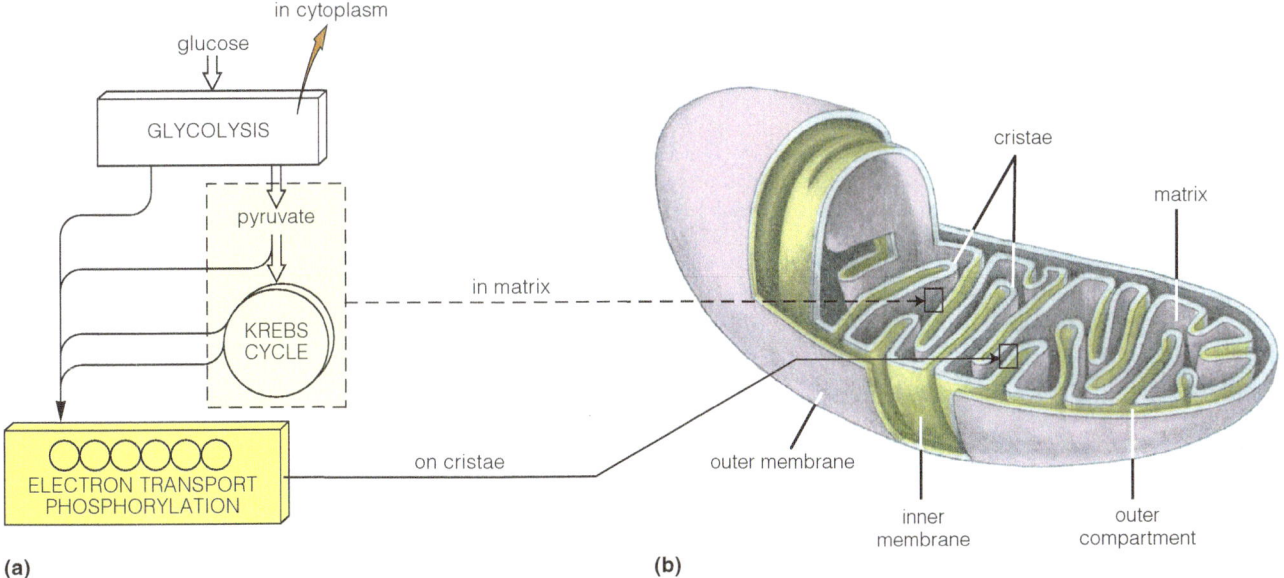

(a)   (b)

Figure 10-5 (a) The pathways in aerobic respiration. (b) The membranes and compartments of a mitochondrion.

2. Now observe Figure 10-6, a high-magnification electron micrograph of a mitochondrion. Identify and label *the outer membrane* separating the organelle from the cytoplasm.

3. Note the presence of an inner membrane, folded into fingerlike projections. Each projection is called a *crista* (the plural is *cristae*). The folding of the inner membrane greatly increases the surface area on which many of the chemical reactions of aerobic respiration take place. Label a crista.

4. Identify and label the *outer compartment,* the space between the inner and outer membranes. The outer compartment serves as a reservoir for hydrogen ions.

5. Finally, identify and label the *inner compartment* (filled with the *matrix*), the interior of the mitochondrion.

6. Study Figure 10-5a, a diagram of the pathways in aerobic respiration and where in the cell they take place.

Now that you know the structure of the mitochondrion, you can visualize the events that take place to produce the chemical energy needed for life. Glycolysis, the first step in *all* respiration pathways, takes place in the cytoplasm. Pyruvate, a carbohydrate, and energy carriers formed during glycolysis enter the mitochondrion during aerobic respiration, moving through both the outer and inner membranes to the matrix within the inner compartment.

Within the inner compartment, the pyruvate is broken down in the Krebs cycle, forming more energy-carrier molecules as well as CO_2. A small amount of ATP is also produced during these reactions.

Electron transport molecules are embedded on the inner membrane, and ATP production occurs as hydrogen ions cross from the outer compartment to the inner compartment. Water is also formed. This accomplishes the third portion of aerobic respiration, electron transfer phosphorylation.

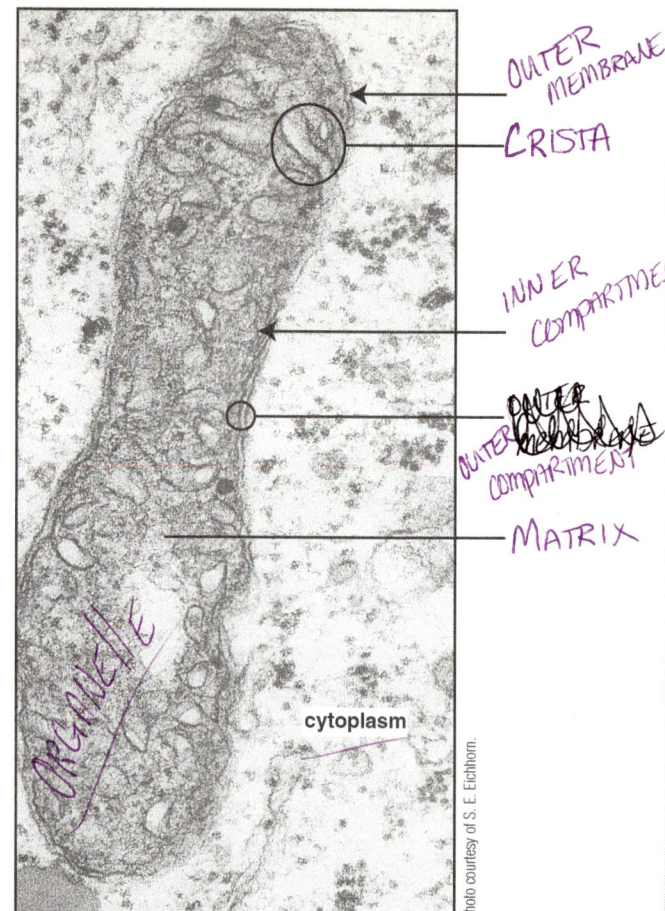

Figure 10-6 Transmission electron micrograph of a mitochondrion (18,600×).
Labels: ~~outer membrane~~, ~~inner compartment~~, ~~crista~~, ~~outer compartment~~, ~~matrix~~

Note: After completing all labs, take your dirty glassware to the sink and wash it following the directions given in *"Instructions for Washing Laboratory Glassware,"* page x. Invert the test tubes in the test tube racks so that they drain. Tidy up your work area, making certain all equipment used in this exercise is there for the next class.

1. The energy-releasing process that occurs in most organisms when oxygen is present is
(a) alcoholic fermentation.
(b) aerobic fermentation.
(c) aerobic respiration.
(d) adenosine triphosphate.

 2. The "universal energy currency" of the cell is
(a) O_2.
(b) $C_6H_{12}O_6$.
(c) ATP.
(d) H_2O.

 3. Products of aerobic respiration include
(a) glucose.
(b) oxygen.
(c) carbon dioxide.
(d) starch.

 4. "Efficiency" of a respiration pathway refers to the
(a) number of steps in the pathway.
(b) amount of CO_2 produced relative to the amount of carbohydrate entering the pathway.
(c) amount of H_2O produced relative to the amount of carbohydrate entering the pathway.
(d) amount of ATP energy produced relative to the energy content of the carbohydrate entering the pathway.

 5. Phenol red is used in Part 10.1.A. to visualize
(a) O_2 production from aerobic respiration.
(b) CO_2 production from aerobic respiration.
(c) sugar production from aerobic respiration.
(d) O_2 consumption in aerobic respiration.

 6. Tetrazolium chloride is used by the seed industry to indicate
(a) whether a seed is undergoing photosynthesis.
(b) if seeds are undergoing aerobic respiration.
(c) if seeds are alive.
(d) both b and c

 7. Which of these is a direct source of energy for muscle contraction?
(a) ATP
(b) glycogen
(c) glucose
(d) fatty acids

 8. Bromthymol blue (BTB) is _____-colored below pH 6.0 and _____-colored above pH 7.6.
(a) blue, yellow
(b) yellow, green
(c) red, blue
(d) yellow, blue

 9. Yeast cells undergoing alcoholic fermentation produce
(a) glucose.
(b) ethanol.
(c) H_2O.
(d) all of the above

 10. Alcoholic fermentation is the basis for
(a) bread baking.
(b) beer brewing.
(c) ethanol fuel production.
(d) all of the above

EXERCISE **10**

Respiration: Energy Conversion

10.1 Aerobic Respiration

10.1.A. Experiment: Carbon Dioxide Production

1. Explain the role of the following components in the experiment on carbon dioxide production:

   germinating corn seeds _____

   ungerminated (dry) corn seeds _____

   germinating, boiled corn seeds _____

   phenol red solution _____

2. Cyanide affects the electron transport system of mitochondria by binding to components of the system and inhibiting the transfer of electrons to oxygen. If seeds were treated with cyanide, what results would they show in the tetrazolium test? Explain.

10.1.B. Effect of Physical Activity on Cellular Respiration

3. Describe at least two reasons why the time it takes your breath to produce a color change in BTB after running 1 minute might be different from the time recorded for another person?

4. As physical activity ramps up, a person typically begins to breathe more rapidly, to sweat, and to flush red. Describe how each of these physical responses of the body is connected to the reactants, products, and byproducts of the process of aerobic respiration.

10.2 Experiment: Alcoholic Fermentation and Ethanol Production

5. Sucrose (table sugar) is a disaccharide composed of glucose and fructose. Glycogen is a polysaccharide composed of many glucose subunits. Which of the following fermentation tubes would you expect to produce the greatest gas volume over a 1-hour period? Why?

   Tube 1: glucose plus yeast

   Tube 2: sucrose plus yeast

   Tube 3: glycogen plus yeast

6. Bread is made by mixing flour, water, sugar, and yeast to form a dense dough. Why does the dough rise? What gas is responsible for the holes in bread?

10.3 Ultrastructure of the Mitochondrion

7. Examine the artificially-colored electron micrograph of the mitochondrion on the right.

   (a) What portions of aerobic respiration occur in region A?

   _____

   (b) What substance is produced as hydrogen ions cross from the space between the inner and outer membranes?

   _____

   (c) What portion(s) of cellular respiration takes(s) place in the cytoplasm *outside* of this organelle?

   _____

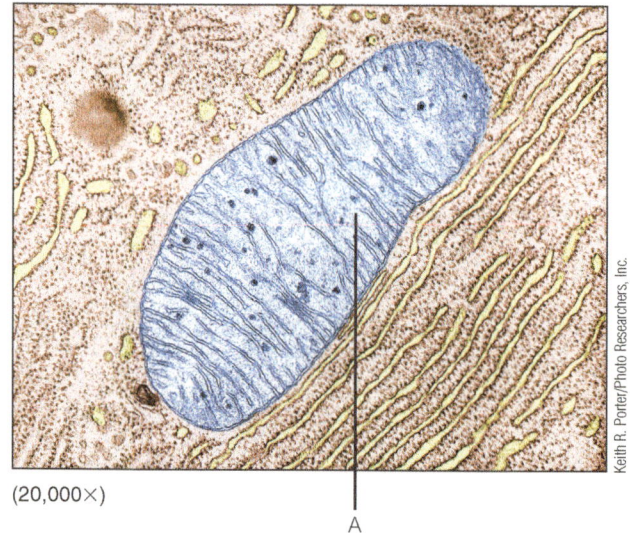

(20,000×)

A

Keith R. Porter/Photo Researchers, Inc.

Food for Thought

8. Compare aerobic respiration and fermentation in terms of

   (a) efficiency of obtaining energy from glucose

   (b) end products

9. The first law of thermodynamics seems to conflict with what we know about ourselves. For example, after strenuous exercise we run out of "energy." We must eat to replenish our energy stores. Where has that energy gone? What form has it taken?

10. Government subsidy programs for corn ethanol stimulated great production increases in the U.S. in the early years of the twenty-first century. The programs were controversial due to disagreements over a number of issues, including the efficiency of energy production from ethanol relative to that required to grow corn crop and transport and manufacture it, as well as the economic effects caused by diverting a food crop to energy production (more than half of U.S. corn grain is fed to livestock, and many ingredients of processed foods are derived from corn). Search two Internet sites for information on benefits and drawbacks of producing ethanol from corn in the United States. List the two sites below and briefly summarize what you learn from each.

   http:// _____

   _____

   _____

   _____

   http:// _____

   _____

   _____

   _____

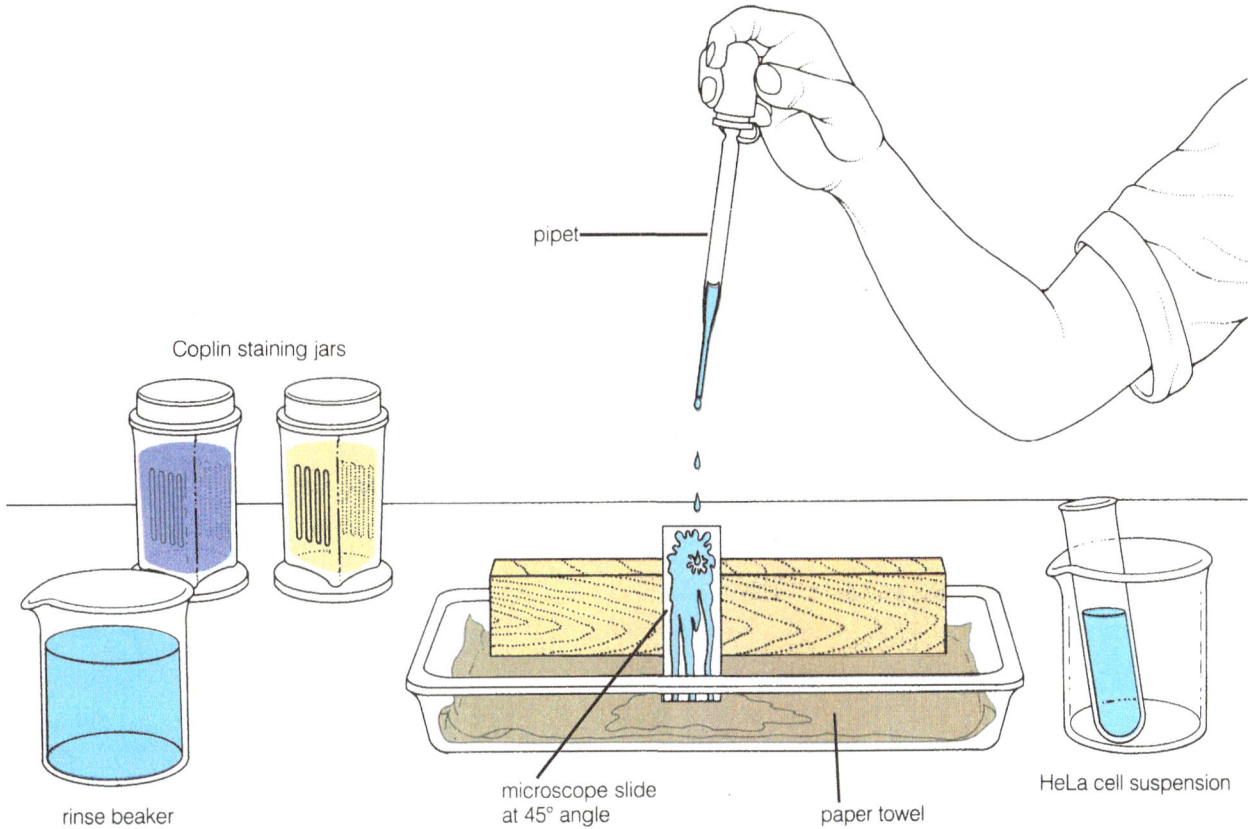

Figure 11-2 Applying HeLa cells to a slide.

5. Go to the location of the staining solutions and dip the slide in stain 1 for 1 second. Withdraw the slide and then dip twice more. Drain the slide of the excess stain, blotting the bottom edge of the slide on the paper toweling before proceeding.

6. Immediately dip the slide in stain 2 for 1 second. (Repeat twice more.) Drain and blot the excess stain as before.

7. Rinse the slide in dH$_2$O by swishing it gently back and forth in the beaker.

8. Allow the slide to *air-dry completely.*

9. Add a drop of water onto the preparation and make a traditional wet mount slide.

10. Place the slide on the microscope stage and observe your chromosome spread, focusing first with the medium-power objective and then with the high-dry objective. Locate cells that appear to have burst and have the chromosomes spread out (see Figure 11-3.) The number of good spreads will be low; so careful observation of many cells is necessary.

Note: **If your microscope has an oil-immersion objective, proceed to step 11. If not, skip to step 12.**

11. Once a good spread has been located, rotate the high-dry objective out of the light path and place a drop of immersion oil on that spot. Rotate the oil-immersion objective into the light path. Be very careful to focus only with the fine-focus of the microscope.

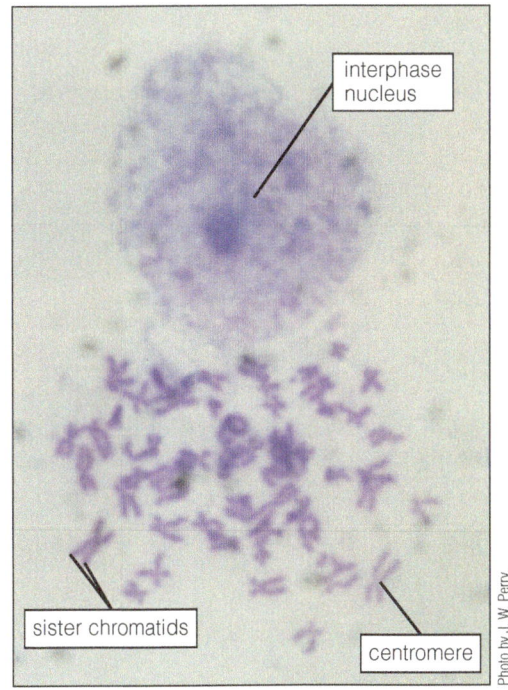

Figure 11-3 Stained chromosome spread of HeLa cells. Sister chromatids connected at their centromeres are clearly visible (w.m., 1000×).

12. Observe the structure of the chromosomes, identifying sister chromatids and centromeres. Draw several of the chromosomes in Figure 11-4, labeling these parts.

13. Select 10 cells in which the chromosomes are visible and count the number of chromosomes you observe in each cell, recording your data in Table 11-1. When you have counted the chromosomes in 10 individual cells, calculate the average number of chromosomes per cell.

Now examine Figure 11-5, an electron micrograph of a human chromosome. Label the two chromatids, the centromere, and the duplicated chromosome.

Note: **Use illustrations in your textbook to aid you in the following study.**

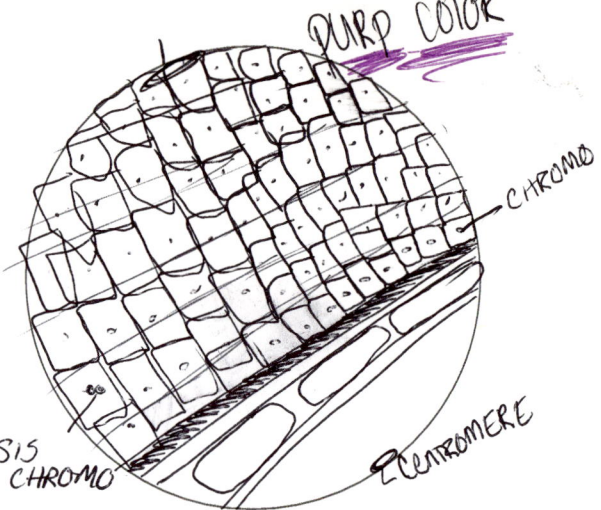

Figure 11-4 Drawing of human chromosomes from HeLa cells (_40_ ×).
Labels: chromosome, sister chromatid, centromere

TABLE 11-1	Chromosome Numbers in HeLa Cells											
Cell	1	2	3	4	5	6	7	8	9	10	Average
Number of chromosomes											

© Cengage Learning 2013

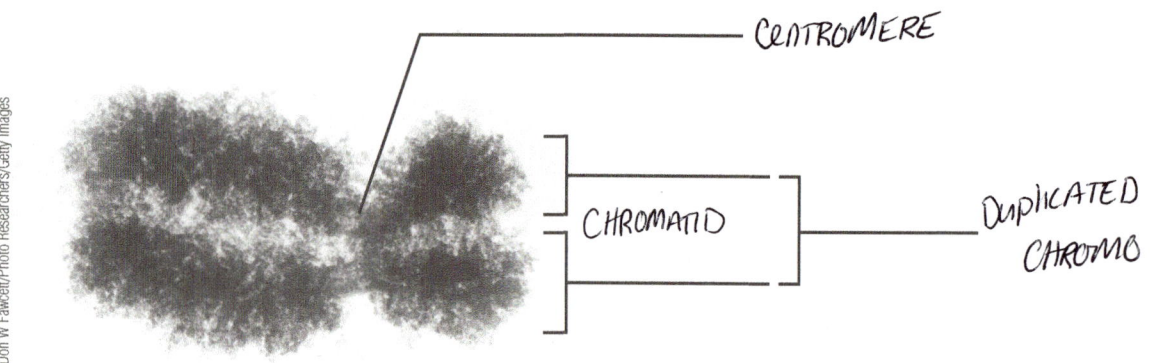

Don W Fawcett/Photo Researchers/Getty Images

Figure 11-5 Electron micrograph of a chromosome (26,000×).
Labels: chromatid (2), centromere, duplicated chromosome

11.2 The Cell Cycle in Plant Cells: Broad Bean Root Tip Squash
(About 45 min.)

Nuclear and cell divisions in plants are, for the most part, localized in specialized regions called meristems. Meristems are regions of active growth. A meristem contains cells that have the capability to divide repeatedly.

Plants have two types of meristems: apical and lateral. Apical meristems are found at the tips of plant organs (shoots and roots) and increase length. Lateral meristems, located beneath the bark of woody plants, increase girth. In this section, you will examine structures related to the cell cycle in the apical meristem of broad bean roots. The tips of actively growing roots may have been prestained for you, or your instructor may provide you with instructions for staining them yourself.

MATERIALS

Per student:

- clean microscope slides
- forceps
- pencil with eraser
- compound microscope
- razor blade or scalpel (optional)

Per student group:

- beaker with stained root tips of broad bean
- dropper bottle of 45% acetic acid
- Kimwipes™ or other tissues

PROCEDURE

1. Obtain a single stained root of broad bean and place it on a clean slide. With a second slide or razor blade, cut off all but the stained tip (approximately 1–3 mm). Discard the remainder and keep the stained tip.
2. Add a drop of 45% acetic acid onto the stained root tip and add a coverslip. Try to prevent air bubbles in your preparation. Allow the root tip to remain in the acetic acid for one or two minutes.
3. The root tip preparation process softened the tissue so that it can be spread out into a single layer of cells. Tap the root tip area with a pencil eraser until the spot is much paler in color and has spread to 8 to 10 mm in diameter. Try not to move the coverslip; if it moves, it will cause some of the cells to roll and fold.

> **CAUTION**
>
> 45% acetic acid can cause burns. Rinse your hands well with water if you get acid on them.

4. With a folded Kimwipe or tissue on top of the coverslip, use the ball of your thumb or the heel of your hand to press down firmly, absorbing excess acetic acid into the tissue and flattening the root tip.
5. Observe the squash preparation with low magnification. If the cells are in a single layer with little overlapping, continue. If most of the cells are in two or more layers, continue tapping with the pencil until the squash is light pink and well spread.
6. Look for cuboidal cells that are actively dividing under low, and then medium power. Use high power to scrutinize cells closely. Locate and study cells in all the phases of the cell cycle, such as in Figure 11-6. Keep in mind that the cell cycle is a continuous cycle; we separate events into different stages as a convenience to aid in their study, but it is often difficult to say definitively when one phase begins and another ends.

A. Interphase and Mitosis

1. **Interphase**. Use the medium-power objective to scan the squash. Note that most of the nuclei are in interphase.

    Switch to the high-dry objective, focusing on a single interphase cell. Note the distinct **nucleus**, with one or more **nucleoli**, and the **chromatin** dispersed within the bounds of the **nuclear envelope**. Label these features in cell 1 of Figure 11-7.
2. **Mitosis**
    (a) *Prophase*. During **prophase** the chromatin folds and condenses; the duplicated chromosomes become visible as threadlike structures. At the same time, microtubules outside the nucleus are beginning to assemble into **spindle fibers**. Collectively, the spindle fibers make up the **spindle,** a three-dimensional structure widest in the middle and tapering to a point at the two *poles* (opposite ends of the cell). *You will not see the spindle during prophase.*

    Find a nucleus in prophase. Draw and label a prophase nucleus in cell 2 of Figure 11-7.

    The transition from prophase to metaphase is marked by the fragmentation and disappearance of the nuclear envelope. At about the same time, the nucleoli also disappear.

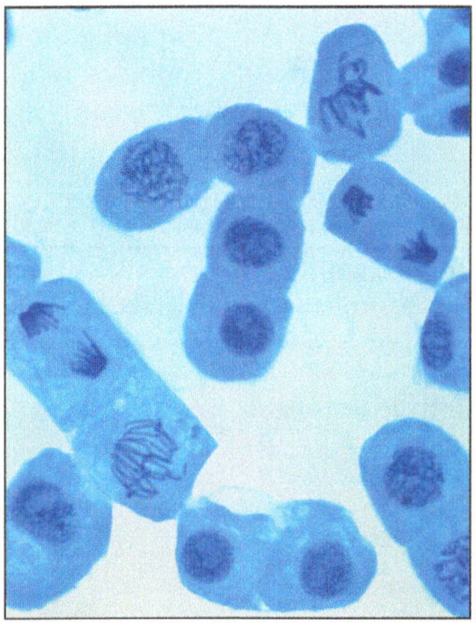

    (b) *Metaphase*. When the nuclear envelope is no longer distinct, the cell is in metaphase. Identify a metaphase cell by locating a cell with the duplicated chromosomes, each consisting of two **sister chromatids,** lined up midway between the two poles. This imaginary midline is called the **spindle equator**. (You may not be able to distinguish the chromatids.) The spindle has moved into the space the nucleus once occupied. The microtubules have become attached to the chromosomes at the *kinetochores,* groups of proteins that form the outer faces of the centromeres (most chromosomes of broad bean have the centromere near one end). Find a cell in metaphase. Label cell 3 of Figure 11-7.

Figure 11-6 Broad bean root tip squash.

Spike Walker/Stone/Getty Images

(c) *Anaphase.* During **anaphase**, sister chromatids of each chromosome separate, each chromatid moving toward an opposite pole. Find an early anaphase cell, recognizable by the slightly separated chromatids. Notice that the chromatids begin separating at the centromere. The last point of contact before separation is complete is at the ends of the "arms" of each chromatid. Although incompletely understood, the mechanism of chromatid separation is based on action of the spindle-fiber microtubules. Once separated, each chromatid is referred to as an individual daughter chromosome. Note that now the chromosome consists of a *single* chromatid. Find a late anaphase cell and draw it in cell 4 of Figure 11-7.

(d) *Telophase.* When the daughter chromosomes arrive at opposite poles, the cell is in **telophase**. The spindle disorganizes. The chromosomes expand again into chromatin form, and a nuclear envelope re-forms around each newly formed daughter nucleus. Find a telophase cell and label individual chromosomes, nuclei, and nuclear envelopes on cell 5 of Figure 11-7.

B. Cytokinesis in Broad Bean Cells

Cytokinesis, division of the cytoplasm, usually follows mitosis. In fact, it often overlaps with telophase. Find a cell undergoing cytokinesis in the broad bean root tip. In plants, cytokinesis takes place by **cell plate formation** (Figure 11-8). During this process, Golgi body–derived vesicles filled with cellulose and other cell wall materials migrate to the spindle equator, where they fuse. Their contents contribute to the formation of a new cell wall, and their membranes make up the new plasma membranes. In most plants, cell plate formation starts in the *middle* of the cell.

1. Examine Figure 11-8, an electron micrograph showing cell plate formation. Note the microtubules that are part of the spindle apparatus.
2. Find a cell undergoing cytokinesis on the prepared slide of broad bean root tips. With your light microscope, the developing **cell plate** appears as a line running horizontally between the two newly formed nuclei. Return to cell 5 of Figure 11-7 and label the developing cell plate.

   Recently divided cells are often easy to distinguish by their square, boxy appearance. Find two recently divided **daughter cells**; then draw and label their contents in cell 6 of Figure 11-7. Include cytoplasm, nuclei, nucleoli, nuclear envelopes, and chromatin.

   Following cytokinesis, the cell undergoes a period of growth and enlargement, during which time it is in interphase. Interphase may be followed by another mitosis and cytokinesis, or in some cells interphase may persist for the rest of a cell's life.

C. Duration of Phases of the Cell Cycle

A broad bean root tip preparation can be thought of as a snapshot capturing cells in various phases of the cell cycle at a particular moment in time. The frequency of occurrence of a cell cycle phase is directly proportional to the length of a phase. You can therefore estimate the amount of time each phase takes by tallying the proportions of cells in each phase.

The length of the cell cycle for cells in actively dividing broad bean root tips is approximately 24 hours, with mitosis lasting for about 90 minutes.

1. Examine a region of the broad bean root tip squash slide where cells in mitosis are abundant. Count the number of cells in each of the stages of mitosis plus interphase in one field of view. Repeat this procedure for other fields of view until you count 100 cells. Record your data in Table 11-2.

TABLE 11-2 Determining Duration of Cell Cycle Phases			
Phase	**Number Observed**	**% of Total**	**Duration (hrs)**
Interphase			
Prophase			
Metaphase			
Anaphase			
Telophase			
Total			

© Cengage Learning 2013

2. Now calculate the time spent in each stage based on a 24-hour cell cycle by dividing the number of cells in each stage by the total number of cells counted to determine percent of total cells in each stage.
3. Multiply the fraction obtained in step 2 by 24 to determine duration.

Although your calculations are only a rough approximation of the time spent in each stage, they do illustrate the differences in duration of each stage in the cell cycle.

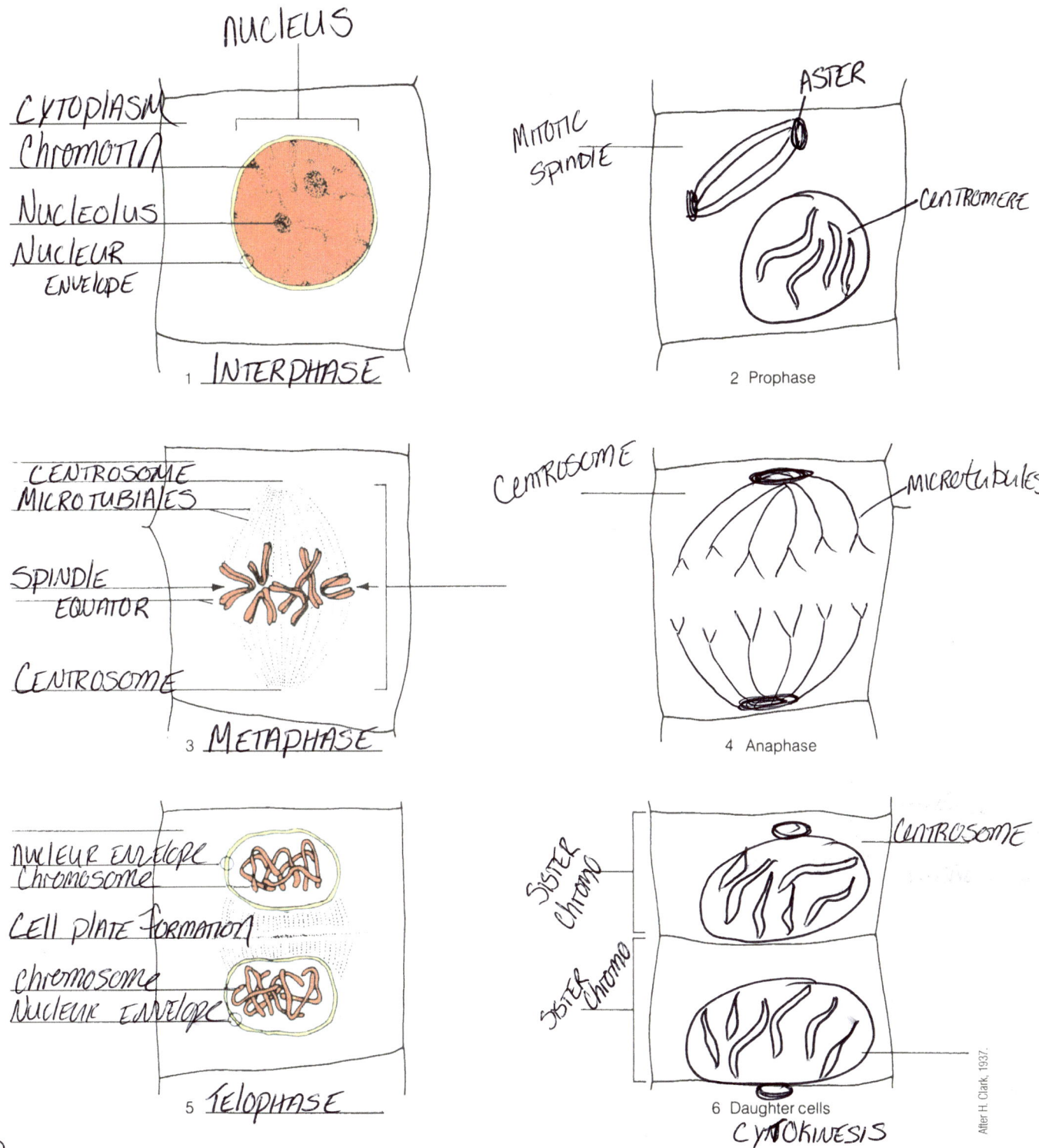

Figure 11-7 Interphase, mitosis, and cytokinesis in broad bean cells.
Labels: interphase, cytoplasm, nucleus, nucleolus, chromatin, nuclear envelope, metaphase, spindle fibers, spindle, pole, spindle equator (between arrows), sister chromatids, telophase and cell plate formation, chromosome, cell plate, daughter cell
(*NOTE:* Some terms are used more than once.)

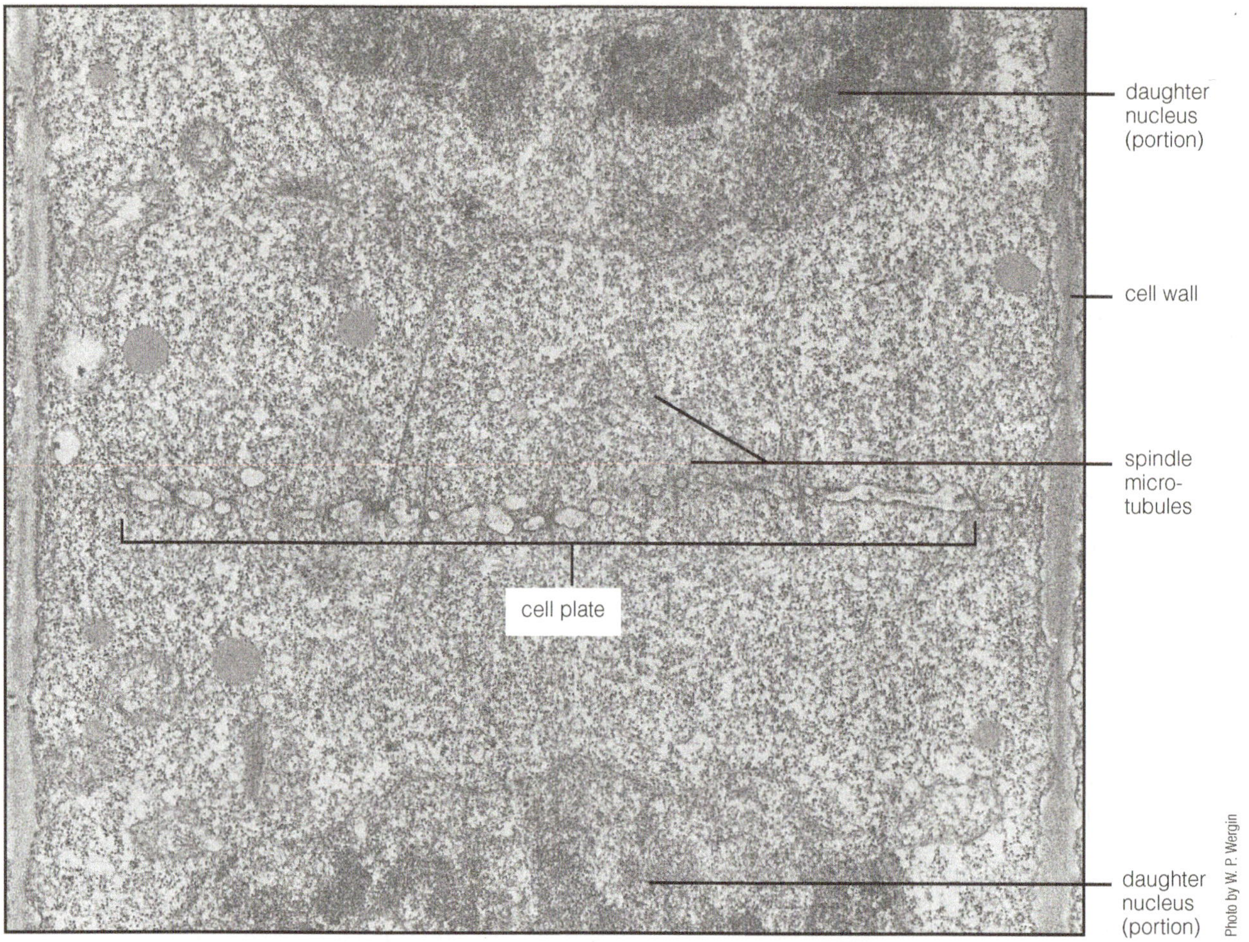

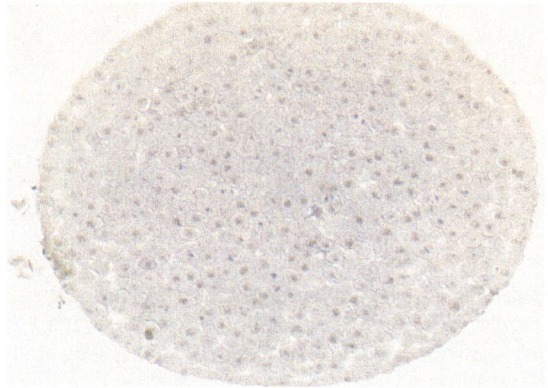

daughter nucleus (portion)

cell wall

spindle microtubules

cell plate

daughter nucleus (portion)

Photo by W. P. Wergin

Figure 11-8 Transmission electron micrograph of cytokinesis by cell plate formation in a plant cell (2000×).

11.3 The Cell Cycle in Animal Cells: Whitefish Blastula (*About 20 min.*)

Fertilization of an ovum by a sperm produces a zygote. In animal cells, the zygote undergoes a special type of cell division, *cleavage*, in which no increase in cytoplasm occurs between divisions. A ball of cells called a *blastula* is produced by cleavage. Within the blastula, repeated nuclear and cytoplasmic divisions take place; consequently, the whitefish blastula is an excellent example in which to observe the cell cycle of an animal.

Note a key difference between plants and animals: whereas plants have meristems where divisions continually take place, animals do not have specialized regions to which mitosis and cytokinesis are limited. Indeed, divisions occur continually throughout many tissues of an animal's body, replacing worn-out or damaged cells.

With several important exceptions, mitosis in animals is remarkably like that in plants. These exceptions will be pointed out as we go through the cell cycle in the whitefish blastula.

MATERIALS

Per student:

- prepared slide of whitefish blastula mitosis
- compound microscope

PROCEDURE

Obtain a slide labeled "whitefish blastula." Scan it with the low-power objective and then at medium power. This slide has numerous thin sections of a blastula. Select one section (Figure 11-9) and then switch to the high-dry objective for detailed observation.

As you examine the slides, draw the cells to show the correct sequence of events in the cell cycle of whitefish blastula.

Photo by J. W. Perry.

Figure 11-9 Section of a whitefish blastula (75×).

A. Interphase and Mitosis

1. **Interphase.** Locate a cell in **interphase.** As you observed in the broad bean root tip, note the presence of the nucleus and chromatin within it. Note also the absence of a cell wall.

   Draw an interphase cell above the word "Interphase" in Figure 11-10 and label the cytoplasm, nucleus, and plasma membrane.

2. **Mitosis**

   (a) *Prophase.* The first obvious difference between mitosis in plants and animals is found in **prophase.**

   Unlike the broad bean cells, those of whitefish contain **centrioles** (Figure 11-11). As seen with the electron microscope, centrioles are barrel-shaped structures consisting of nine radially arranged triplets of microtubules.

   One pair of centrioles was present in the cytoplasm in the G1 stage of interphase. These centrioles duplicated during the S stage of interphase. Subsequently, one new and one old centriole migrated to each pole.

   Although the centrioles are too small to be resolved with your light microscope, you can see a starburst pattern of spindle fibers that appear to radiate from the centrioles. Other microtubules extend between the centrioles, forming the **spindle** (Figure 11-12). The chromosomes become visible as the chromatin condenses.

   Find a prophase cell, identifying the spindle and starburst cluster of fibers about the centriole.

   Draw the prophase cell in the proper location on Figure 11-10. Label the spindle, chromosomes, cytoplasm, and the position of the plasma membrane.

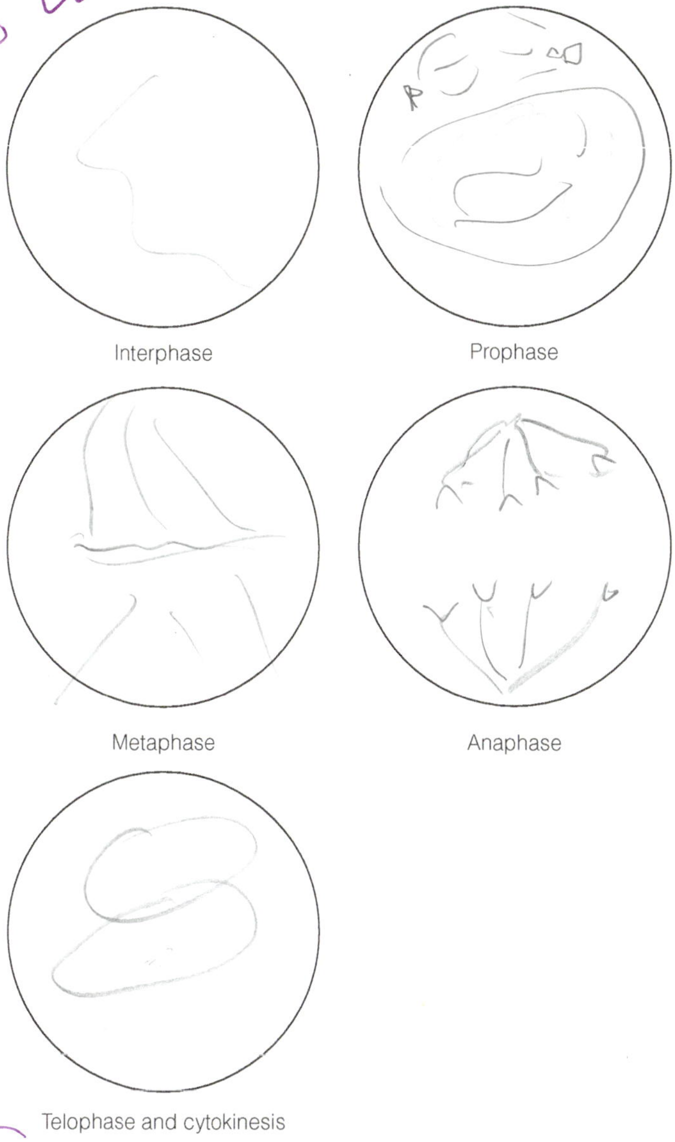

Interphase   Prophase

Metaphase   Anaphase

Telophase and cytokinesis

Figure 11-10   Drawings of cell cycle stages in whitefish blastula. **Labels:** cytoplasm, nucleus, plasma membrane, spindle, chromosomes, spindle equator, sister chromatids, daughter nuclei, chromatin, furrow.
(*NOTE:* Some terms are used more than once.)

   (b) *Metaphase.* As was the case in plant cells, during **metaphase** the spindle fiber microtubules become attached to the *kinetechore* of each centromere region, and the duplicated chromosomes (each consisting of two **sister chromatids**) line up on the **spindle equator**. Locate a metaphase cell.

   Draw the metaphase cell in the proper location on Figure 11-10. Label the chromosomes on the spindle equator, spindle, and plasma membrane.

   (c) *Anaphase.* Again similar to that observed in plant cells, **anaphase** begins with the separation of sister chromatids into individual (daughter) chromosomes. Observe a blastula cell in anaphase. Draw the anaphase cell in the proper location on Figure 11-10. Label the separating sister chromatids, spindle, cytoplasm, and plasma membrane.

   (d) *Telophase.* **Telophase** is characterized by the arrival of the individual (daughter) chromosomes at the poles. A nuclear envelope forms around each daughter nucleus. Find a telophase cell.

   Is the spindle still visible? _NO_

   Is there any evidence of a nuclear envelope forming around the chromosomes? _YES_

   Draw the telophase cell in Figure 11-10. Label daughter nuclei, chromatin, cytoplasm, and plasma membrane.

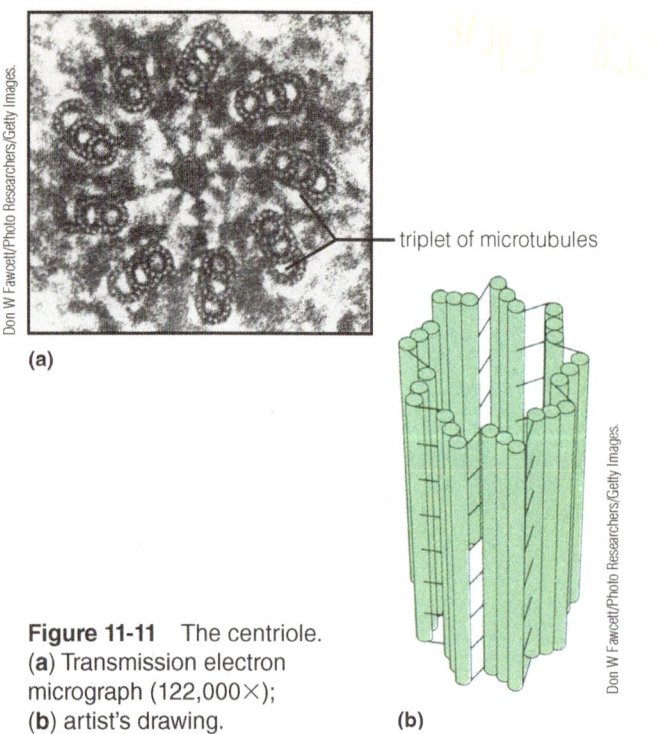

triplet of microtubules

Figure 11-11 The centriole.
(a) Transmission electron micrograph (122,000×);
(b) artist's drawing.

(a)

(b)

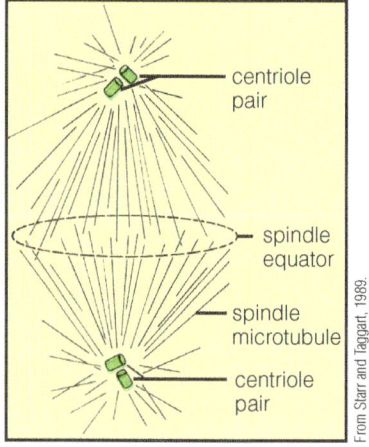

centriole pair

spindle equator

spindle microtubule

centriole pair

Figure 11-12 Spindle apparatus in animal cell.

B. Cytokinesis in Animal Cells

A second major distinction between cell division in plants and animals occurs during cytoplasmic division. Cell plates are absent in animal cells. Instead, cytokinesis takes place by **furrowing**.

To visualize how furrowing takes place, imagine wrapping a string around a balloon and slowly tightening the string until the balloon has been pinched in two. In life, the animal cell is pinched in two, forming two discrete cytoplasmic entities, each with a single nucleus. Figure 11-13 illustrates the cleavage furrow in an animal cell.

Find a cell in the blastula undergoing cytokinesis. The telophase cell that you drew in Figure 11-10 may also show an early stage of cytokinesis. Label the cleavage furrow if it does.

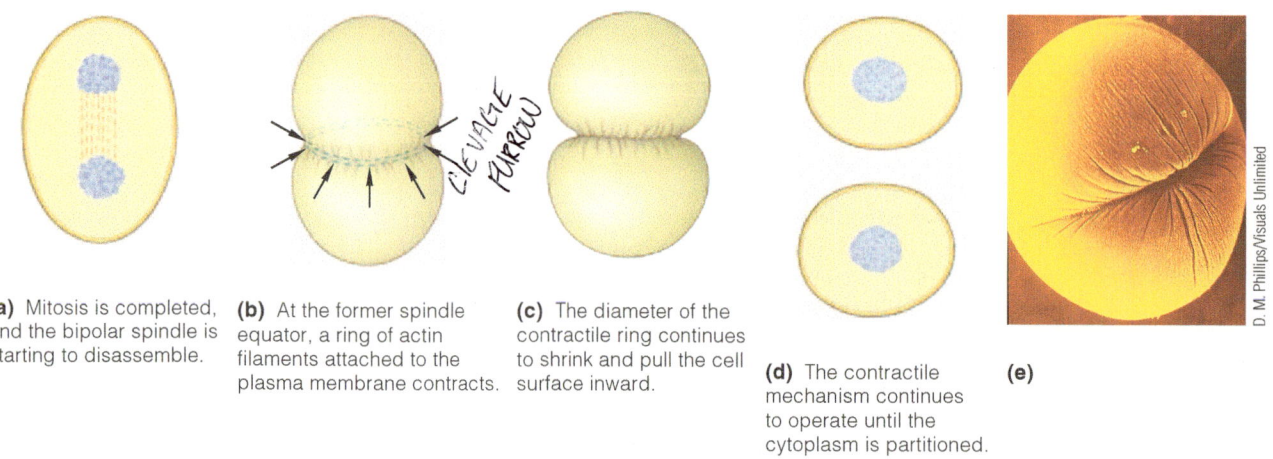

CLEVAGE FURROW

(a) Mitosis is completed, and the bipolar spindle is starting to disassemble.

(b) At the former spindle equator, a ring of actin filaments attached to the plasma membrane contracts.

(c) The diameter of the contractile ring continues to shrink and pull the cell surface inward.

(d) The contractile mechanism continues to operate until the cytoplasm is partitioned.

(e)

Figure 11-13 **(a–d)** Cytoplasmic division of an animal cell. **(e)** Scanning electron micrograph of the cleavage furrow.

Understanding chromosome movements is crucial to understanding mitosis. You can simulate mitosis with a variety of materials. This is a simple activity, but a valuable one. It will be especially helpful when comparing the events of mitosis with those of meiosis in the next exercise.

MATERIALS

Per student pair:
- 44 pop beads each of two colors
- eight magnetic centromeres

PROCEDURE

1. Build the components for two pairs of chromosomes by assembling strings of pop beads as follows:
   (a) Assemble two strands of pop beads with eight pop beads of one color on each arm, with a magnetic centromere connecting the two arms.
   (b) Repeat step a, but use pop beads of the second color.
   (c) Assemble two more strands of pop beads, but with three pop beads of one color on each arm.
   (d) Repeat step c, using pop beads of the second color.

   You should have four long strings, two of each color, and four short strings, with two of each color. Each pop bead string should have a magnetic centromere at its midpoint. Note that pop bead strings can attach to each other at the magnetic centromere. Each pop bead string represents a single molecule of DNA plus proteins.

2. Place **one** of each kind of strand in the center of your workspace, which represents the nucleus. You have created a nucleus with four "chromosomes," two long and two short.

3. Manipulate these model chromosomes through the phases of the cell cycle, beginning in the G1 phase of interphase and proceeding through the rest of interphase, mitosis, and cytokinesis.

4. Check with your instructor to be sure that you are modeling the process correctly.

D 1. Reproduction in prokaryotes occurs primarily through the process known as
(a) mitosis.
(b) cytokinesis.
(c) furrowing.
(d) fission.

C 2. The genetic material (DNA) of eukaryotes is organized into
(a) centrioles.
(b) spindles.
(c) chromosomes.
(d) microtubules.

B 3. The process of cytoplasmic division is known as
(a) meiosis.
(b) cytokinesis.
(c) mitosis.
(d) fission.

B 4. The product of chromosome duplication is
(a) two chromatids.
(b) two nuclei.
(c) two daughter cells.
(d) two spindles.

A 5. The correct sequence of stages in *mitosis* is
(a) interphase, prophase, metaphase, anaphase, telophase.
(b) prophase, metaphase, anaphase, telophase.
(c) metaphase, anaphase, prophase, telophase.
(d) prophase, telophase, anaphase, interphase.

A 6. During prophase, duplicated chromosomes
(a) consist of chromatids.
(b) contain centromeres.
(c) consist of nucleoproteins.
(d) contain all of the above.

D 7. During the S period of interphase,
(a) cell growth takes place.
(b) nothing occurs because this is a resting period.
(c) chromosomes divide.
(d) synthesis (or replication) of the nucleoproteins takes place.

D 8. Chromatids separate during
(a) prophase.
(b) telophase.
(c) cytokinesis.
(d) anaphase.

A 9. Cell plate formation
(a) occurs in plant cells but not in animal cells.
(b) usually begins during telophase.
(c) is a result of fusion of Golgi vesicles.
(d) is all of the above.

C 10. Centrioles and a starburst cluster of spindle fibers would be found in
(a) both plant and animal cells.
(b) only plant cells.
(c) only animal cells.
(d) none of the above

EXERCISE **11**

Mitosis and Cytokinesis: Nuclear and Cytoplasmic Division

Post-Lab Questions

Introduction

1. Distinguish among interphase, mitosis, and cytokinesis.

11.1 Chromosomal Structure

2. Distinguish between the structure of a duplicated chromosome before mitosis and the chromosome produced by separation of two chromatids during mitosis.

11.2 The Cell Cycle in Plant Cells: Broad Bean Root Tip Squash

3. If the chromosome number of a typical broad bean root tip cell is 12 before mitosis, what is the chromosome number of each newly formed nucleus after mitosis has taken place?

4. In plants, what name is given to a region where mitosis occurs most frequently?

5. The plant cells in the following photomicrographs have been stained to show microtubules comprising the spindle apparatus. Identify the stage of mitosis in each and label the region indicated on **(b)**.

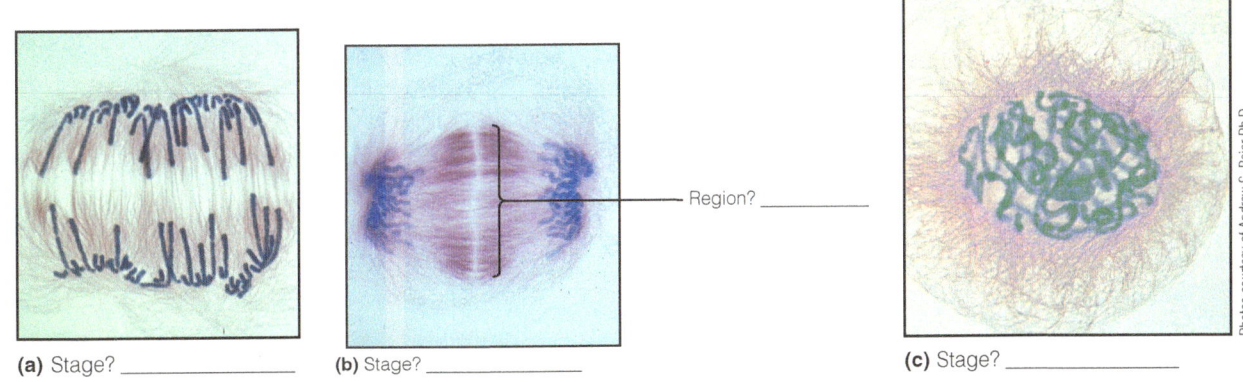

— Region? _____

Photos courtesy of Andrew S. Bajer Ph.D.

(a) Stage? _____   **(b)** Stage? _____   **(c)** Stage? _____

6. Name two features of animal cell mitosis and cytokinesis you can use to distinguish these processes from those occurring in plant cells.

   (a)

   (b)

Food for Thought

7. Observe photomicrographs (a) and (b) below. Note the double nature of the blue "threads" in (a). Each individual component of the doublet is called a(an) _____.
   Is (b) an image of a plant or an animal cell? How do you know?

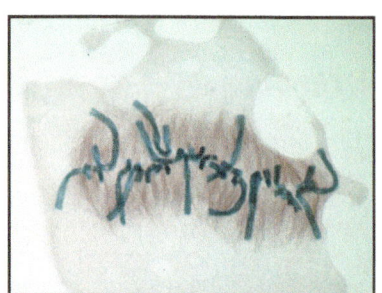

(a) Name of structure? _____

Photos courtesy of Andrew S. Bajer Ph.D

(b) Plant or animal? _____
   How do you know? _____

8. Why do you suppose cytokinesis generally occurs in the cell's midplane?

9. Why must the DNA be duplicated during the S phase of the cell cycle, prior to mitosis?

10. Tumor cells arise from normal body cells after a series of cellular "mistakes" disrupts the controls over, and processes of, mitosis and cytokinesis. Describe two errors that might happen during the processes of mitosis and/or cytokinesis and that would produce a daughter cell with "extra" chromosomes.

Meiosis: Basis of Sexual Reproduction

OBJECTIVES

After completing this exercise, you will be able to

1. define *meiosis, homologue (homologous chromosome), gene, diploid, haploid, gamete, ovum, sperm, fertilization, zygote, spore, allele, gene pair, locus, synapsis, genotype, crossing over, nonlinked, linked, nondisjunction;*

2. describe differences and similarities between meiosis and mitosis;

3. describe the basic similarities and differences between the life cycles of higher plants and higher animals;

4. describe the process of meiosis, and recognize events that occur during each stage;

5. discuss the significance of crossing over, segregation, and independent assortment;

6. describe the process of nondisjunction and how it produces chromosome number abnormalities in resulting gametes and zygotes.

Introduction

Mitosis alone is not adequate for every nuclear division job required to complete a life cycle. Sexual reproduction requires **meiosis**, which, like mitosis, is a process of nuclear division. During mitosis, the number of chromosomes in the daughter nuclei remains the same as that in the parental nucleus. In meiosis, however, the genetic complement is halved, resulting in daughter nuclei containing only one-half the number of chromosomes as the parental nucleus. Thus, while mitosis is sometimes referred to as an *equational division*, meiosis is often called *reduction division*. Moreover, while mitosis is completed after a single nuclear division, two divisions, called meiosis I and meiosis II, occur during meiosis. Table 12-1 summarizes differences between mitosis and meiosis.

TABLE 12-1	**Comparison of Mitosis and Meiosis**
Mitosis	**Meiosis**
Equational division: amount of genetic material remains constant.	Reduction division: amount of genetic material is halved.
Completed in one division cycle.	Requires two division cycles for completion.
Produces two genetically identical nuclei.	Produces two to four genetically different nuclei.
Generally produces cells not directly involved in sexual reproduction.	Ultimately produces cells used for sexual reproduction.

© Cengage Learning 2013

In the body cells of most eukaryotes, chromosomes exist in pairs called **homologues** (homologous chromosomes); that is, there are two chromosomes that are physically similar and contain the same **genes**, which are sections of DNA that are units of inheritance.

When both homologues are in the *same* nucleus, the nucleus is **diploid** (2n); when only one of the homologues is present, the nucleus is **haploid** (n). A parental nucleus normally contains the diploid (2n) chromosome number before meiosis; all four daughter nuclei contain the haploid (n) number at the completion of meiosis.

The reduction in chromosome number is the basis for sexual reproduction. In animals, the cells containing the daughter nuclei produced by meiosis are called **gametes**: **ova** (singular is *ovum*) if the parent is female, **sperm** cells if male. As you probably know, gametes are produced in the gonads—ovaries and testes, respectively. In fact, this is the *only* place where meiosis occurs in higher animals. Figure 12-1 shows where meiosis occurs in humans, and Figure 12-2 shows where and when meiosis occurs during the life cycle of a higher animal.

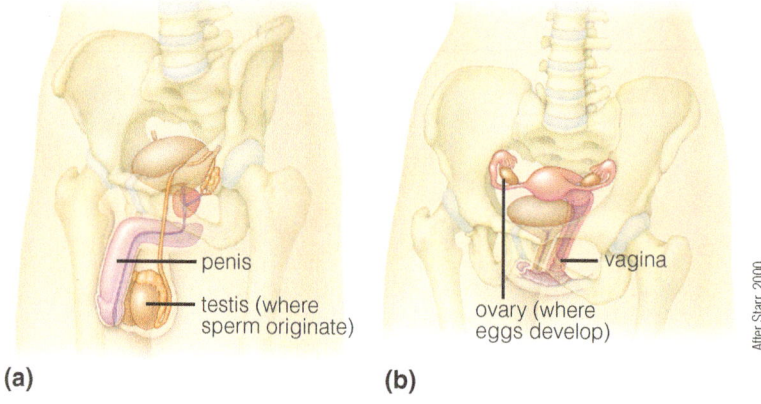

Figure 12-1 Gamete-producing structures in humans: (**a**) human male, (**b**) human female.

Note when meiosis occurs—during gamete production. During **fertilization** (the fusion of a sperm nucleus with an ovum nucleus), the diploid chromosome number is restored as the two haploid gamete nuclei fuse to form the **zygote**, the first cell of the new diploid generation.

What about plants? Do plants have sex? Indeed they do. However, the plant life cycle is a bit more complex than that of animals. Plants of a single species have two completely different body forms. The primary function of one is the production of gametes. This plant is called a *gametophyte* ("gamete-producing plant") and all of its cells are haploid. Because the entire plant is haploid, gametes are produced in specialized organs by mitosis. The other body form, a *sporophyte*, is diploid. This diploid sporophyte has specialized organs in which meiosis occurs, producing haploid spores (hence the name *sporophyte*, "spore-producing plant"). When spores germinate and produce more cells by mitosis, they grow into haploid gametophytes, completing the life cycle.

Figure 12-3 shows the structures in a typical flower that produce sperm and eggs.

Examine Figure 12-4, which shows the gametophyte and sporophyte of a fern plant. Remember, the gametophyte and sporophyte are different, free-living stages of the *same* species of fern. Now look at Figure 12-5, which diagrams a typical plant life cycle. Again, note the consequence of meiosis. In plants, it results in the production of **spores**, not gametes.

You should understand an important concept from these diagrams: *Meiosis always halves the chromosome number. The diploid chromosome number is eventually restored when two haploid nuclei fuse during fertilization.*

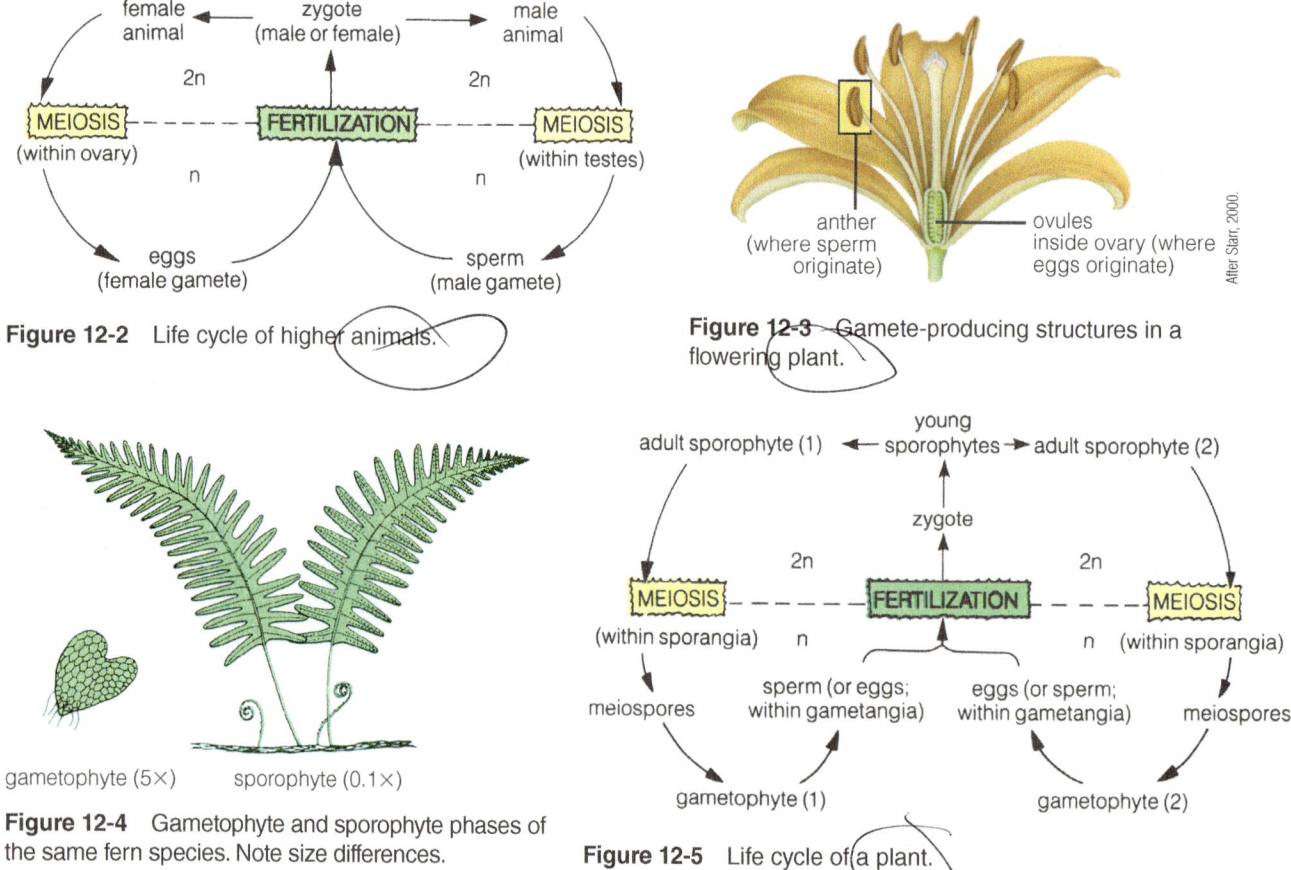

Figure 12-2 Life cycle of higher animals.

Figure 12-3 Gamete-producing structures in a flowering plant.

Figure 12-4 Gametophyte and sporophyte phases of the same fern species. Note size differences.

Figure 12-5 Life cycle of a plant.

Understanding meiosis is an absolute necessity for understanding the patterns of inheritance in Mendelian genetics, the subject of Exercise 13. Gregor Mendel, an Austrian monk, spent years deciphering the complexity of simple genetics. Although he knew nothing of genes and chromosomes, he noted certain patterns of inheritance and formulated three principles, now known as Mendel's principles of recombination, segregation, and independent assortment. The following activities will demonstrate the events of meiosis and the genetic basis for Mendel's principles.

12.1   Demonstrations of Meiosis Using Pop Beads *(About 60 min.)*

MATERIALS

Per student pair:

- 44 pop beads each of two colors (red and yellow, for example), as used in Exercise 11
- eight magnetic centromeres
- marking pens
- eight pieces of string, each 40 cm long
- meiotic diagram cards similar to those used here
- colored pencils

Per student group (table):

- bottle of 95% ethanol to remove marking ink
- tissues

PROCEDURE

Work in pairs.

Within the nucleus of an organism, each chromosome bears the genes, the units of inheritance. Genes may exist in two or more alternative forms called **alleles**. Each homologue bears *genes* for the same traits; these are the **gene pairs**. However, the homologues may or may not have the same *alleles*.

An example will help here: suppose the trait in question is flower color and that a flower has only two possible colors, red or white (Figure 12-6a, b). The gene, designated with a letter "R," is coding (providing the information) for flower color. There are two homologues in the same nucleus, so each bears the gene for flower color. *But*, on one homologue, the *allele* might code for red flowers ("R"), while the allele on the other homologue might code for white flowers ("r," Figure 12-6c). There are two other possibilities. The alleles on *both* homologues might be coding for red flowers (Figure 12-6d), or they *both* might be coding for white flowers (Figure 12-6e). Note that these three possibilities are mutually exclusive in a nucleus.

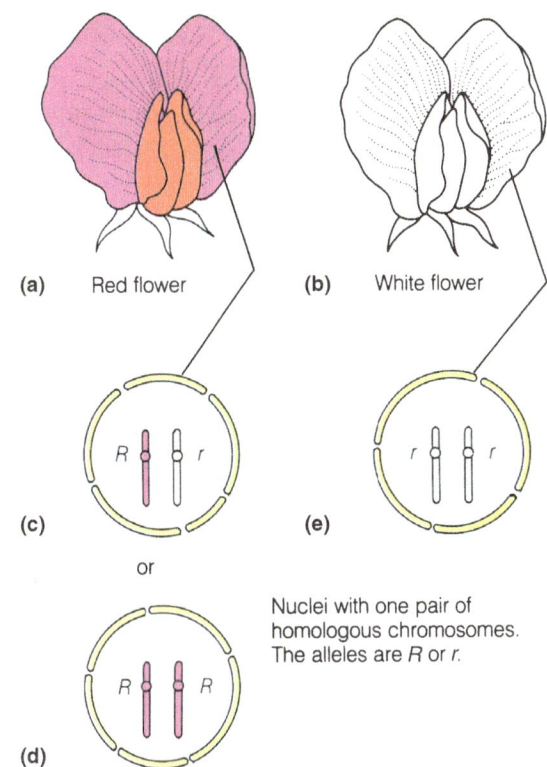

Figure 12-6 Chromosomal control of flower color. (c–e) show a nucleus with one pair of homologous chromosomes. The alleles for flower color are *R* or *r*.

1. Build the components for two pairs of homologous chromosomes by assembling strings of pop beads as follows:
   (a) Assemble two strands of pop beads with eight pop beads of one color on each arm, with a magnetic centromere connecting the two arms.
   (b) Repeat step a, but use pop beads of the second color.
   (c) Assemble two more strands of pop beads, but with three pop beads of one color on each arm.
   (d) Repeat step c, using pop beads of the second color.
      You should have four long strings, two of each color, and two short strings, also two of each color. Each pop-bead string should have a magnetic centromere at its midpoint, by which pop-bead strings can attach to each other. Each pop-bead string represents a single molecule of DNA plus proteins, with each bead representing a gene.
2. Place **one** of each kind of strand in the center of your workspace, which represents the interphase nucleus of a cell that will undergo meiosis. You have created a nucleus with four "chromosomes," two long and two short. The long strands represent one homologous pair, and the short strands represent a second homologous pair of chromosomes.

We start by assuming that these chromosomes represent the diploid condition. The two colors represent the origin of the chromosomes: one homologue (color: _____YELLOW_____) came from the male parent, and the other homologue (color: _____PINK_____) came from the female parent.

3. The four single-stranded chromosomes represent four unduplicated chromosomes. Now simulate DNA duplication during the S-phase of interphase (Figure 11-1), when each DNA molecule and its associated proteins are copied exactly. The two copies, called sister chromatids, remain attached to each other at their centromeres (Figure 12-7). During chromosome replication, the genes also duplicate. Thus, alleles on sister chromatids are identical.

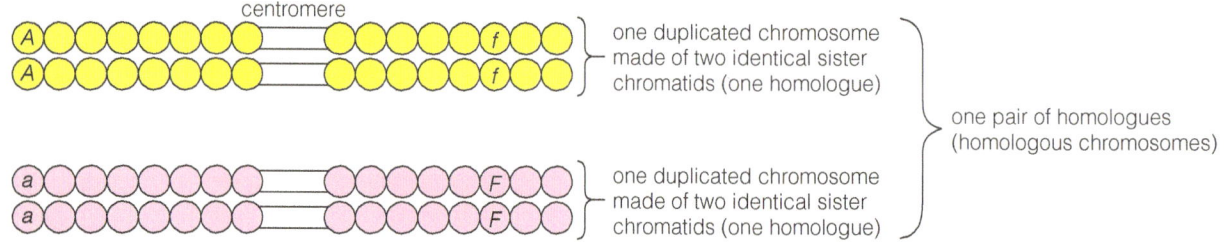

Figure 12-7 One pair of homologous chromosomes.

How many sister chromatids are there in a duplicated chromosome? ___2___

How many chromosomes are represented by four sister chromatids? ___2___ By eight? ___16___

What is the diploid number of the starting (parental) nucleus you've created? (*Hint:* Count the number of homologues to obtain the diploid number.) ___32___

4. As mentioned previously, genes may exist in two or more alternative forms, called alleles. The location of an allele on a chromosome is its **locus** (plural: *loci*). Using the marking pen, mark two loci on each long chromatid with letters to indicate alleles for a common trait. Suppose the long pair of homologous chromosomes codes for two traits, skin pigmentation and the presence of attached earlobes in humans. We'll let the capital letter *A* represent the allele for normal pigmentation and a lowercase *a* the allele for albinism (the absence of skin pigmentation); *F* will represent free earlobes and *f* attached earlobes. A suggested marking sequence is illustrated in Figure 12-7.

5. Let's assign a gene to our second homologous pair of chromosomes, the short pair. We'll suppose this gene codes for the production of an enzyme necessary for metabolism. On one homologue (consisting of two chromatids) mark the letter *E*, representing the allele causing enzyme production. On the other homologue, *e* represents the allele that interferes with normal enzyme production.

6. Obtain a meiotic diagram card like the one in Figure 12-8. Manipulate your model chromosomes through the stages of meiosis described below, moving the chromosomes to the correct diagram circles (representing nuclei) as you go along. Reference to Figure 12-8 will be made at the proper steps. *DO NOT* draw on the meiotic diagram cards.

A. Meiosis without Crossing Over

Although crossing over is a nearly universal event during meiosis, we will first work with a simplified model to illustrate chromosomal movements and separations during meiosis. Refer to Figure 12-9 as you manipulate your model.

1. **Late interphase.** During interphase, the nuclear envelope is intact and the chromosomes are randomly distributed throughout the nucleoplasm (semifluid substance within the nucleus). All duplicated chromosomes (eight chromatids) should be in the parental nucleus, indicating that DNA duplication has taken place. The sister chromatids of each homologue should be attached by their magnetic centromeres, but the four homologues should be separate. Your model nucleus contains a diploid number 2n = 4.

   The pop-bead chromosomes should appear during interphase in the parental nucleus as shown in Figure 12-8. Be sure to mark the location of the alleles. Use different pencil or pen colors to differentiate the homologues on your drawings.

2. **Meiosis I.** During meiosis I, homologues are separated from each other into different nuclei. Daughter nuclei created are thus haploid.

   (a) *Prophase I.* During the first prophase, the parental nucleus contains four duplicated homologous chromosomes, each comprised of two sister chromatids joined at their centromeres. The chromatin condenses to form discrete, visible chromosomes. The homologues pair with each other. This pairing is called **synapsis.** Slide the two homologues together.

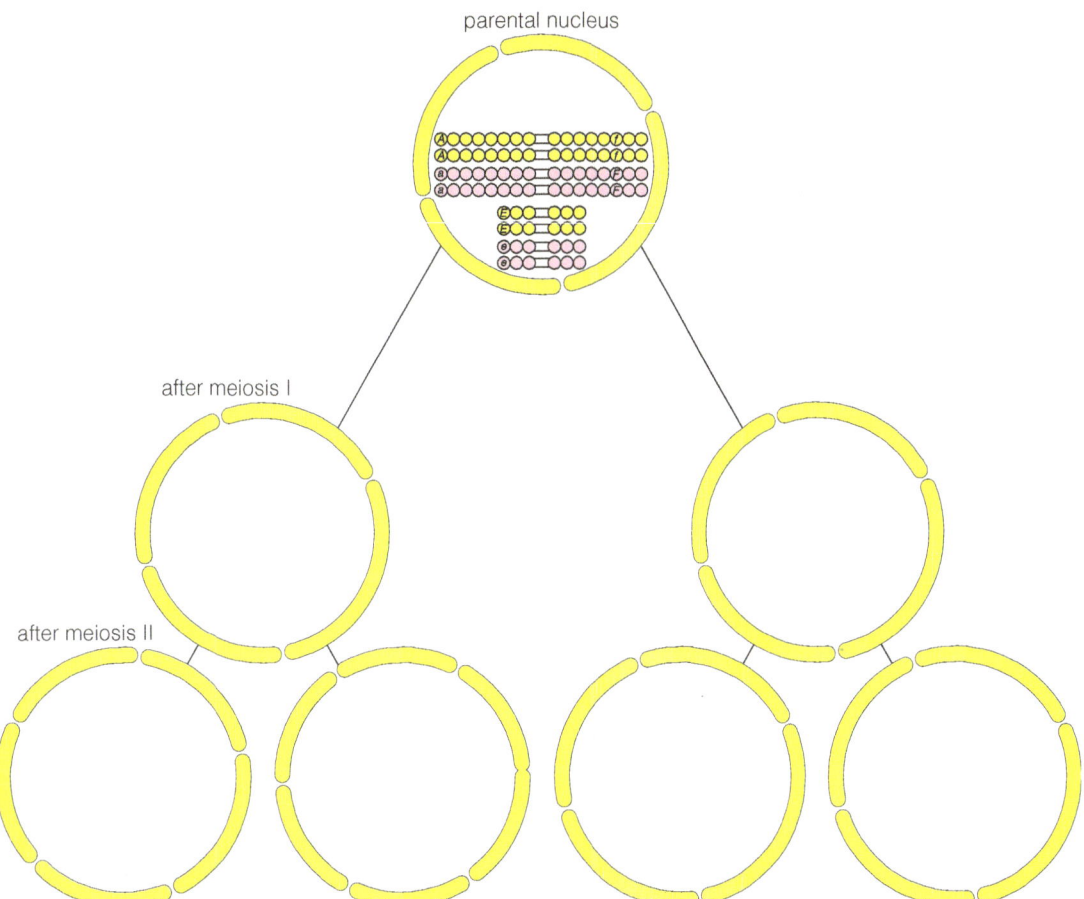

Figure 12-8 Meiosis without crossing over.

Twist the chromatids about one another to simulate synapsis.

The nuclear envelope disorganizes at the end of prophase I.

(b) *Metaphase I.* Homologous chromosomes now move toward the spindle equator, the centromeres of each homologue coming to lie *on either side of the equator.* Spindle fibers, consisting of aggregations of microtubules, attach to the centromeres. One homologue attaches to microtubules extending from one pole, and the other homologue attaches to microtubules extending from the opposite spindle pole.

To simulate the spindle fibers, attach one piece of string to each centromere. Then lay the free ends of strings from two homologues toward one spindle pole and the ends of the other homologues toward the opposite pole.

(c) *Anaphase I.* During anaphase I, the homologous chromosomes separate, one homologue moving toward one pole, the other toward the opposite pole. The movement of the chromosomes is apparently the result of shortening of some spindle fibers and lengthening of others. Each homologue is still in the duplicated form, consisting of two sister chromatids.

Pull the two strings of one homologous pair toward its spindle pole and the other toward the opposite spindle pole, separating the homologues from one another. Repeat with the second pair of homologues.

(d) *Telophase I.* Continue pulling the string spindle fibers until each homologue is now at its respective pole. The first meiotic division is now complete. You should have two nuclei, each containing two chromosomes (one long and one short) consisting of two sister chromatids.

Draw your pop-bead chromosomes as they appear after meiosis I on the two nuclei labeled "after meiosis I" of Figure 12-8. Depending on the organism involved, an interphase (interkinesis) and cytokinesis may precede the second meiotic division, *or* each nucleus may enter directly into meiosis II. The chromosomes decondense into chromatin form.

It is important to note here that DNA synthesis *does not* occur following telophase I (between meiosis I and meiosis II).

Before meiosis II, the spindle is rearranged into two spindles, one for each nucleus.

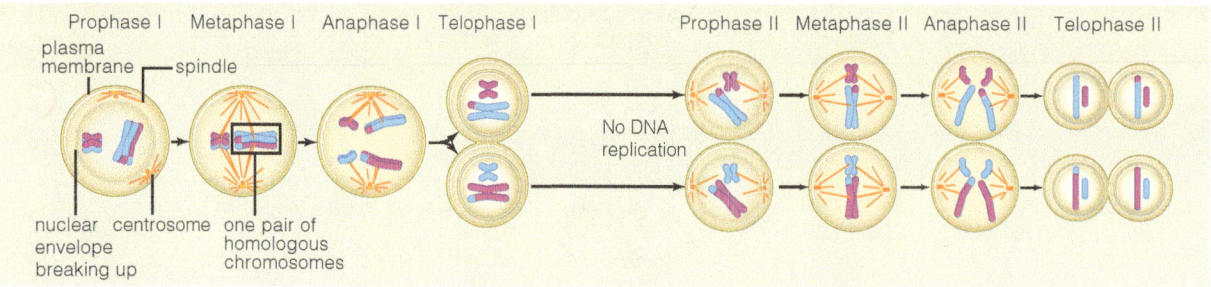

Prophase I Metaphase I Anaphase I Telophase I   Prophase II Metaphase II Anaphase II Telophase II

plasma membrane ┌─ spindle

No DNA replication

nuclear envelope breaking up   centrosome   one pair of homologous chromosomes

Figure 12-9 Meiosis in a generalized diploid germ cell. Two pairs of chromosomes are shown. Maternal chromosomes are shaded *pink*. Paternal chromosomes are shaded *blue*.

3. **Meiosis II.** During meiosis II, sister chromatids are separated into different daughter nuclei. The result is four haploid nuclei.
   (a) *Prophase II.* At the beginning of the second meiotic division, the sister chromatids are still attached by their centromeres. During prophase II, the nuclear envelope disorganizes, and the chromatin recondenses.
   (b) *Metaphase II.* Within each nucleus, the duplicated chromosomes align with the equator, the centromeres lying *on the equator.* Spindle fiber microtubules attach the centromeres of each chromatid to opposite spindle poles.
      Your string spindle fibers should be positioned so that the two spindle fiber strings from sister chromatids lie toward opposite poles. Note that each nucleus contains only *two* duplicated chromosomes (one long and one short) consisting of *two* sister chromatids each.
   (c) *Anaphase II.* The sister chromatids separate, moving to opposite poles. Pull on the string until the two sister chromatids separate. After the sister chromatids separate, each is an individual (not duplicated) daughter chromosome.
   (d) *Telophase II.* Continue pulling on the string spindle fibers until the two daughter chromosomes are at opposite poles. The nuclear envelope reforms around each chromosome and the chromosomes decondense back into chromatin form. Four daughter nuclei now exist. Note that each nucleus contains two individual unduplicated chromosomes (each formerly a chromatid) originally present within the parental nucleus. These nuclei and the cells they're in generally undergo a differentiation and maturation process to become gametes (in animals) or spores (in plants).
      Draw your pop-bead chromosomes as they appear after meiosis II in the "gamete nuclei" of Figure 12-8. Your diagram should indicate the genetic (chromatid) complement *before* meiosis and *after* each meiotic division, *not* the stages of each division.
      Remember that meiosis takes place in both male and female organisms. (See Figure 12-2.)
      If the parental nucleus was from a male, what is the gamete called? _SPERM CELL_
      If female? _OVUM_
      Is the parental nucleus diploid or ~~haploid~~? _Diploid_
      Are the nuclei produced after the *first* meiotic division diploid or ~~haploid~~? _Diploid_
      Are the nuclei of the gametes ~~diploid~~ or haploid? _Haploid_
   { What is the **genotype** of each gamete nucleus after meiosis II? (The genotype is the genetic composition of an organism, or the alleles present. Another way to ask this question is, What alleles are present in each gamete nucleus? Write these in the format: *AFE, afe,* and so on.)

If you answered the preceding questions correctly, you might logically ask, "If the chromosome number of the gametes is the same as that produced after the first meiotic division, why bother to have two separate divisions? After all, the genes present are the same in both gametes and first-division nuclei."

There are two answers to this apparent paradox. The first, and perhaps the most obvious, is that the second meiotic division ensures that a *single* chromatid (nonduplicated chromosome) is contained within each gamete. After gametes fuse, producing a zygote, the genetic material duplicates prior to the zygote's undergoing mitosis. If gametes contained two chromatids, the zygote would have four, and duplication prior to zygote division would produce eight, twice as many as the organism should have. If DNA duplication within the zygote were not necessary for the onset of mitosis, this problem would not exist. Alas, DNA synthesis apparently is a necessity to initiate mitosis.

You can discover the second answer for yourself by continuing with the exercise, for although you have simulated meiosis, you have done so without showing what happens in *real* life. That's the next step.

B. Meiosis with Crossing Over

A very important event that results in a reshuffling of alleles on the chromatids occurs during prophase I. Recall that synapsis results in pairing of the homologues. During synapsis, the chromatids break, and portions of chromatids bearing genes for the same characteristic (but perhaps *different* alleles) are exchanged between *nonsister* chromatids. This event is called **crossing over**, and it results in recombination (shuffling) of the alleles on a chromatid.

1. Look again at Figure 12-7. Distinguish between sister and nonsister chromatids. Now look at Figure 12-10, which demonstrates crossing over in one pair of homologues.
2. Return your chromosome models to the nucleus format with two pairs of homologues entering prophase I.
3. To simulate crossing over, break four beads from the arms of two nonsister chromatids in the long homologue pair, exchanging bead color between the two arms. During actual crossing over, the chromosomes may break anywhere within the arms.

   Crossing over is virtually a universal event in meiosis. Each pair of homologues may cross over in several places simultaneously during prophase I.
4. Manipulate your model chromosomes through meiosis I and II again and watch what happens to the distribution of the alleles as a consequence of the crossing over. Fill in Figure 12-11 as you did before, but this time show the effects of crossing over. Again, use different colors in your sketches.

   What are the genotypes of the gamete nuclei? *Aa Bb*

   Is the distribution of alleles present in the gamete nuclei after crossing over the same as that which was present without crossing over? *—YES*

   Is the distribution of alleles present in the gamete nuclei after crossing over the same as that in the nuclei after the first meiotic division? *—YES*

   Crossing over provides for genetic recombination, resulting in increased variety. How many different genetic *types* of daughter chromosomes are present in the gamete nuclei without crossing over (Figure 12-8)? *TWO*

   How many different types are present with crossing over (Figure 12-11)? *FOUR*

   We think you would agree that a greater number of *types* of daughter chromosomes indicates greater *variety.*

   Recall that the parental nucleus contained a pair of homologues, each homologue consisting of two sister chromatids. Because sister chromatids are identical in all respects, they have the same alleles of a gene (see Figure 12-7). As your models showed, the alleles on nonsister chromatids may not (or may) be identical; they bear the same genes but may have different alleles, different forms of some genes.

   What is the difference between a gene and an allele? *GENE IS THE UNITS OF INHERITANCE, WHILE ALLELE IS A PAIR OF All. Genes.*

   Let's look at the single set of alleles on your model chromosomes that are those for pigmentation, *A* and *a*. Both alleles were present in the parental nucleus. How many are present in the gametes? *Bb*

   This illustrates Mendel's first principle, segregation. Segregation means that during gamete formation, pairs of alleles are separated (segregated) from each other and end up in different gametes.

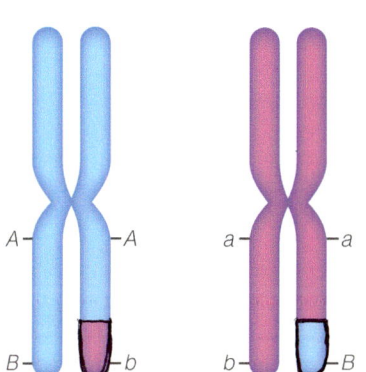

Figure 12-10 Crossing over in one pair of homologues. *Pink* signifies a maternal chromosome, and *blue*, its paternal homologue.

C. Demonstrating Independent Assortment

Manipulate your model chromosomes again through meiosis with crossing over (Figure 12-11), searching for different possibilities in chromosome distribution that would make the gametes genetically different.

Does the distribution of the alleles for enzyme production to different gametes on the second set of homologues have any bearing on the distribution of the alleles on the first set (alleles for skin pigmentation and earlobe condition)? _____

This distribution demonstrates the principle of independent assortment, which states that segregation of alleles into gametes is independent of the segregation of alleles for other traits, *as long as the genes are on*

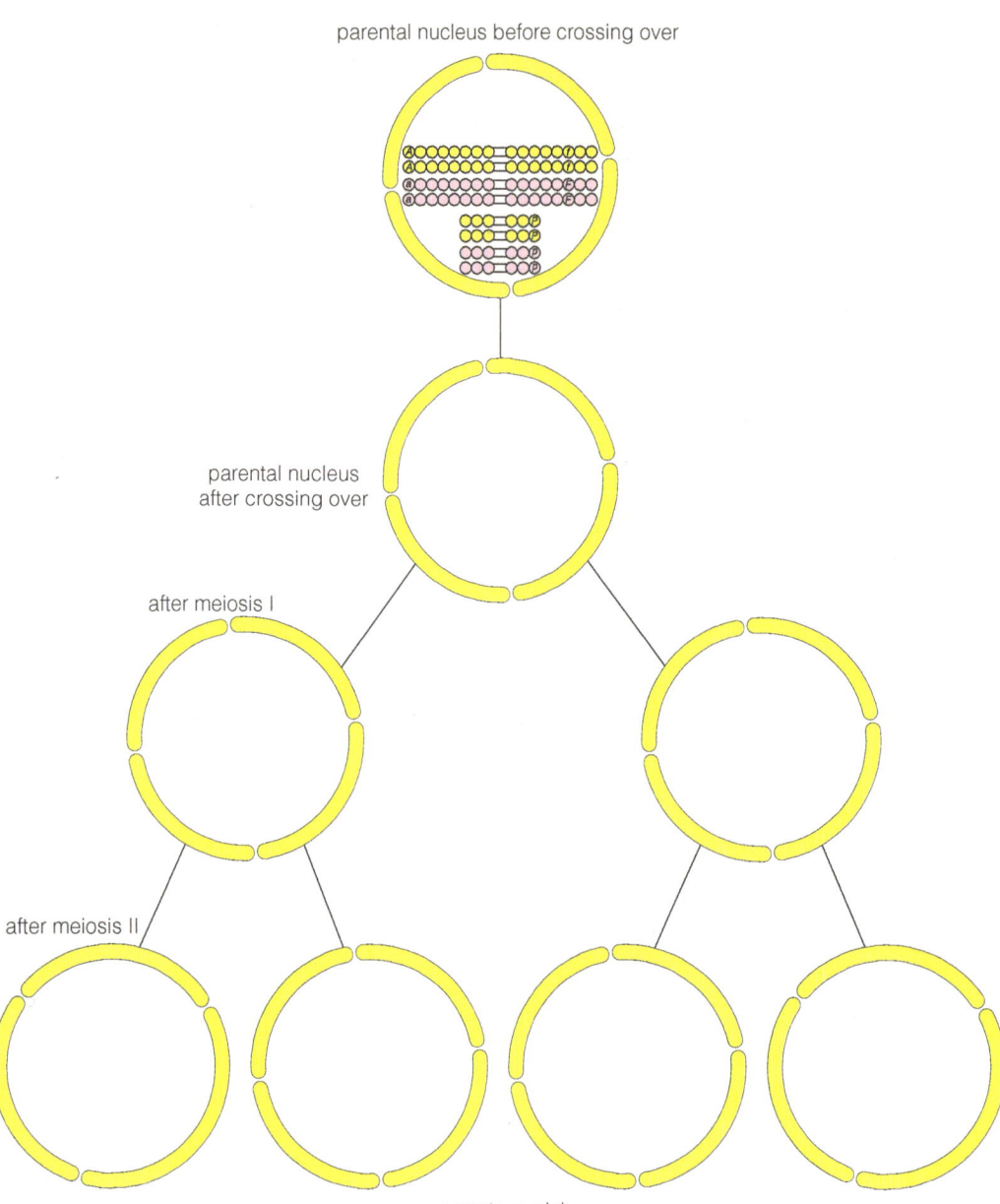

parental nucleus before crossing over

parental nucleus
after crossing over

after meiosis I

after meiosis II

gamete nuclei

Figure 12-11   Meiosis with crossing over.

different sets of homologous chromosomes. Genes that are on *different* (nonhomologous) chromosomes are said to be **nonlinked**. By contrast, genes for different traits that are on the *same* chromosome are **linked**.

Because the genes for enzyme production and those for skin pigmentation and earlobe attachment are on different homologous chromosomes, these genes are *NONLINKED*, while the genes for skin pigmentation and earlobe attachment are *LINKED* because they are on the same chromosome.

In reality, most organisms have many more than two sets of chromosomes. Humans have 23 pairs (2n = 46), while some plants literally have hundreds!

A thorough understanding of meiosis is necessary to understand genetics. With this foundation, you'll find that problems involving Mendelian genetics are easy and fun to do. Without an understanding of meiosis, Mendelian genetics will be hopelessly confusing.

D. Nondisjunction and the Production of Gametes with Abnormal Chromosome Number

Errors in the process of meiosis can occur in many ways. Perhaps the best understood error process is that of **nondisjunction,** when one or more pairs of chromosomes fail to separate in anaphase. The result is gamete nuclei with too few or too many chromosomes.

1. Begin to manipulate your model chromosomes to show meiosis without crossing over (Section 12.1.A). In modeling events at metaphase I, however, arrange the spindle fiber threads for the long pair of homologues so that they all extend to the same pole.
2. Model anaphase I, pulling the chromosomes toward their respective poles. Nondisjunction occurs in the long pair of homologues, with both duplicated chromosomes being pulled to the same pole. See Figure 12-12.

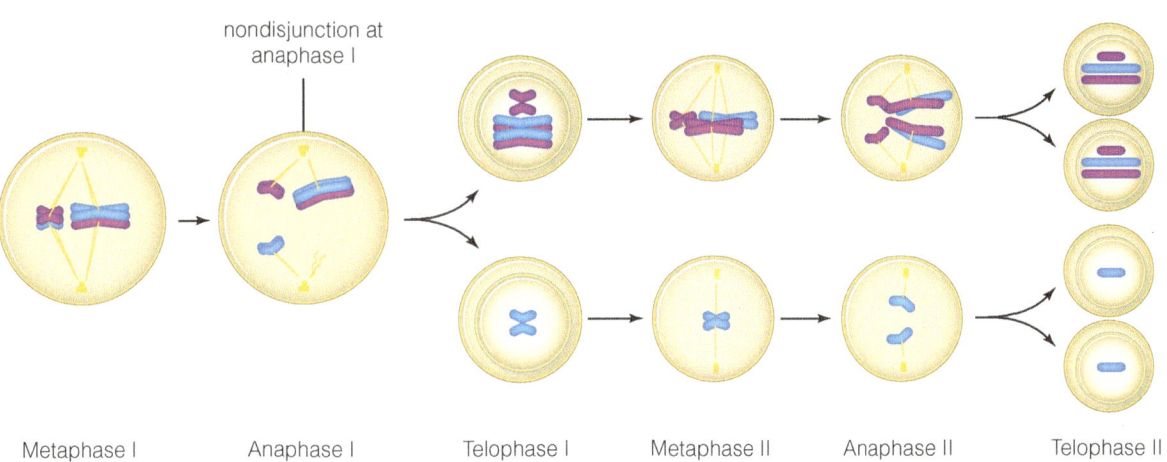

Metaphase I   Anaphase I   Telophase I   Metaphase II   Anaphase II   Telophase II

Figure 12-12 An example of nondisjunction. Of the two pairs of homologous chromosomes shown, one pair fails to separate at anaphase I of meiosis. The chromosome number is altered in the resulting gametes.

3. Continue to manipulate the model chromosomes through the remainder of the meiotic process.
   How many chromosomes are found in gamete nuclei? _____
   How does this compare to the chromosome number in normal gametes? _____
   Recall that each chromosome bears a unique set of genes and speculate about the effect of nondisjunction on the resulting zygotes formed from fertilization with such a gamete.

   _____
   _____
   _____

   One of the most common human genetic disorders arises from nondisjunction during gamete (usually ovum) formation. Down syndrome results from nondisjunction in one of the 23 pairs of human chromosomes, chromosome 21. An individual with Down syndrome has three copies of chromosome 21 instead of the normal two copies. While symptoms of this genetic disorder vary greatly, most individuals show moderate to severe mental impairment and a host of associated physical disorders. Relatively few other human genetic disorders arise from nondisjunction, probably because the consequences of abnormal chromosome number are often lethal.

Note: Remove marking ink from pop beads with 95% ethanol and tissues.

D **1.** In meiosis, the number of chromosomes
_____, while in mitosis, it _____.
(a) is halved/is doubled
(b) is halved/remains the same
(c) is doubled/is halved
(d) remains the same/is halved

A **2.** The term "2n" means
(a) the diploid chromosome number
is present.
(b) the haploid chromosome number
is present.
(c) chromosomes within a single
nucleus exist in homologous pairs.
(d) both a and c

B **3.** In higher animals, including humans,
meiosis results in the production of
(a) egg cells (ova).
(b) gametes.
(c) sperm cells.
(d) all of the above

D **4.** Recombination of alleles on nonsister
chromatids occurs during
(a) anaphase I.
(b) meiosis II.
(c) telophase II.
(d) crossing over.

D **5.** Alternative forms of genes are called
(a) homologues.
(b) locus.
(c) loci.
(d) alleles.

A **6.** If both homologous chromosomes of each
pair exist in the same nucleus, that nucleus is
(a) diploid.
(b) unable to undergo meiosis.
(c) haploid.
(d) none of the above

A **7.** DNA duplication occurs during
(a) interphase.
(b) prophase I.
(c) prophase II.
(d) interkinesis.

D **8.** Nondisjunction
(a) results in gametes with abnormal
chromosome numbers.
(b) occurs at anaphase.
(c) results when homologues fail to
separate properly in meiosis.
(d) is all of the above.

C **9.** The daughter nuclei produced by
meiosis are
(a) genetically identical.
(b) diploid.
(c) haploid.
(d) both a and b

D **10.** Meiosis differs from mitosis in that meiosis
(a) requires two cycles of division for
completion.
(b) produces spores in plants but gametes
in animals.
(c) is only found in animals and not in
plants.
(d) both a and b

EXERCISE **12**

Meiosis: Basis of Sexual Reproduction

Post-Lab Questions

Introduction

1. If a cell of an organism has 46 chromosomes before meiosis, how many chromosomes will exist in each nucleus after meiosis?

2. What basic difference exists between the life cycles of higher plants and higher animals?

3. In animals, meiosis results directly in gamete production, while in plants spores are produced. How are the gametes produced in the life cycle of a plant?

4. How would you argue that meiosis is the basis for sexual reproduction in plants, even though the *direct* result is a spore rather than a gamete?

12.1 Demonstrations of Meiosis Using Pop Beads

5. Suppose one sister chromatid of a chromosome has the allele *H*. What allele will the other sister chromatid have? (Assume crossing over has not taken place.) _____

6. Suppose that two alleles on one homologous chromosome are *A* and *B*, and the other homologous chromosome's alleles are *a* and *b*.
   (a) How many different genetic types of gametes would be produced *without* crossing over? _____
   (b) What are the genotypes of the gametes? _____
   (c) If crossing over were to occur, how many different genetic types of gametes could occur? _____
   (d) List them. _____

7. Assume that you have built a homologous pair of *duplicated* chromosomes, one chromosome red and the other yellow. Describe or draw the appearance of two nonsister chromatids after crossing over.

8. Examine the meiotic diagram at right. Describe in detail what's wrong with it.

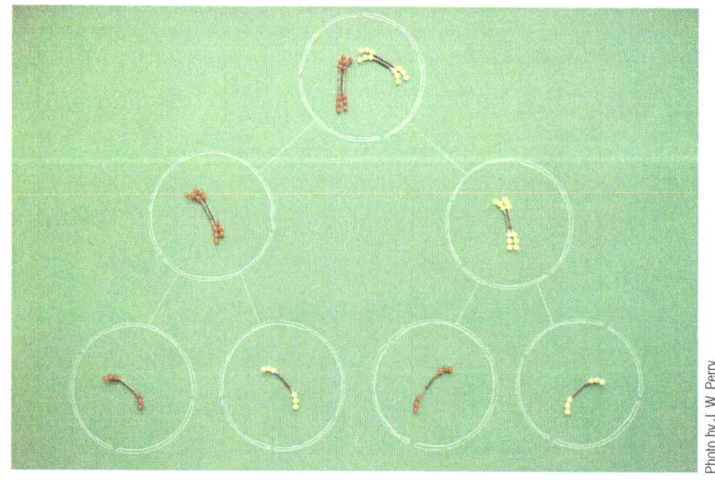

Photo by J. W. Perry.

Food for Thought

9. From a genetic viewpoint, of what significance is fertilization?

10. One of your friends has several brothers and sisters, each quite different in appearance, interests, and abilities, even though all have the same parents. Explain in detail how these siblings reflect crossing over and independent assortment.

Heredity

OBJECTIVES

After completing this exercise, you will be able to

1. define *true-breeding, hybrid, monohybrid cross, law of segregation, diploid, haploid, genotype, phenotype, dominant, recessive, complete dominance, homozygous, heterozygous, probability, chi-square test, dihybrid cross;*

2. solve monohybrid and dihybrid cross problems;

3. use sampling to determine phenotypic ratios of a visible trait in the gametophytes of an F_1 C-fern hybrid;

4. observe sperm release and fertilization events that lead to an F_2 C-fern sporophyte generation;

5. form hypotheses about genotypic and phenotypic ratios in the F_2 C-fern sporophyte generation;

6. use a chi-square test to determine whether observed results are consistent with expected results;

7. determine your phenotype and provide your probable genotype for some common traits.

Introduction

In 1866, an Austrian monk, Gregor Mendel, presented the results of painstaking experiments on the inheritance of the garden pea, but the scientific community ignored them, possibly because they didn't understand their significance. Now, more than a century later, Mendel's work seems elementary to modern-day geneticists, but its importance cannot be overstated. The principles generated by Mendel's pioneering experimentation are the foundation for the genetic counseling so important today to families with genetically based health disorders. They are also the framework for the modern research that is making inroads into treating diseases previously believed to be incurable. In this era of genetic engineering—the incorporation of foreign DNA into chromosomes of other species—it's easy to lose sight of the concepts underlying the processes that make it all possible. These experiments and genetics problems should give you a good basic understanding of these processes.

13.1 Monohybrid Crosses

Garden peas have both male and female parts in the same flower and are able to self-fertilize. For his experiments, Mendel chose parental plants that were **true-breeding,** meaning that all self-fertilized offspring displayed the same form of a trait as their parent. For example, if a true-breeding purple-flowered plant self-fertilizes, all of its offspring will have purple flowers.

When parents that are true-breeding for *different* forms of a trait are crossed—for example, purple flowers and white flowers—the offspring are called **hybrids.** When only one trait is being studied, the cross is a **monohybrid cross.** We'll look first at monohybrid problems and crosses.

A. Monohybrid Problems with Complete Dominance (About 20 min.)

MATERIALS
- simulated chromosomes, consisting of pop beads with magnetic centromeres, and meiotic diagram cards (Page 151, Exercise 12)
- bottle of 70% ethanol

PROCEDURE

1. Most organisms are diploid; that is, they contain homologous chromosomes with genes for the same traits. The location of a gene on a chromosome is its locus (plural: loci). Two genes at homologous loci are called a gene pair. Chromosomes have numerous genes, as shown in Figure 13-1. Genes exist in different forms, called *alleles*. Let's consider one gene pair at the *F* locus. There are three possibilities for the allelic makeup at the *F* locus.

   Both alleles are *FF*:

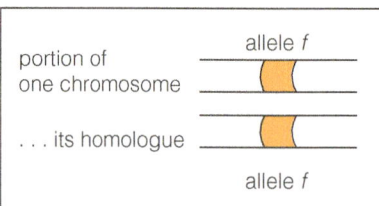

   Both alleles are *ff*:

   One allele is *F*, and the other is *f*:

   Gametes, on the other hand, are **haploid**; they contain only one of the two homologues and thus only one of the two alleles for a specific trait. According to Mendel's first law of inheritance, the **law of segregation**, each organism contains two alleles for each trait, and the alleles segregate (separate) during the formation of gametes during meiosis. Each gamete then contains only one allele of the pair.

   The **genotype** of an organism represents its genetic constitution—that is, the alleles present, either for each locus, or taken cumulatively as the genotype of the entire organism.

   For each of these diploid genotypes, indicate all possible genotypes of the gametes that can be produced by the organism:

Diploid Genotype	Potential Gamete Genotype(s)
FF	*F F*
ff	*ff*
Ff	*Ff* , *Ff*

   In order to review the process that gives rise to the gamete genotypes, manipulate the pop-bead models that you used in Exercise 12. Using a marking pen, label one bead of each chromosome and go through the meiotic divisions that give rise to the gametes. *It is imperative that you understand meiosis before you attempt to do genetics problems.*

2. During fertilization, two haploid gamete nuclei fuse, and the diploid condition is restored. Give the diploid genotype produced by fusion of the following gamete genotypes.

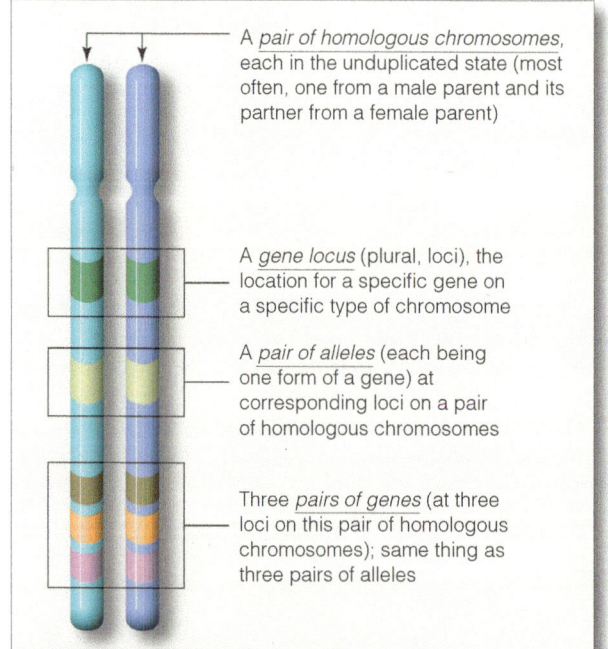

Figure 13-1   A few genetic terms illustrated.

Gamete Genotype	x	Gamete Genotype	⟶	Diploid Genotype
F		F		FF
F		f		Ff
F		f		Ff

3. Now let's attach some meaning to genotypes. As you see, the genotype is the actual genetic makeup of the organism. The **phenotype** is the outward expression of the genotype—that is, what the organism looks like because of its genotype, as well as its physiological traits and behavior. (Although phenotype is determined primarily by genotype, in many instances environmental factors can modify phenotype.)

Human earlobes are either attached or free (Figure 13-2). This trait is determined by a single gene consisting of two alleles, F and f. An individual whose genotype is FF or Ff has free earlobes. This is the **dominant** condition. Note that the presence of one *or* two F alleles results in the dominant phenotype, free earlobes. The allele F is said to be dominant over its allelic partner, f. The **recessive** phenotype, attached earlobes, occurs only when the genotype is ff. In the case of **complete dominance,** the dominant allele completely masks the expression or effect of the recessive allele.

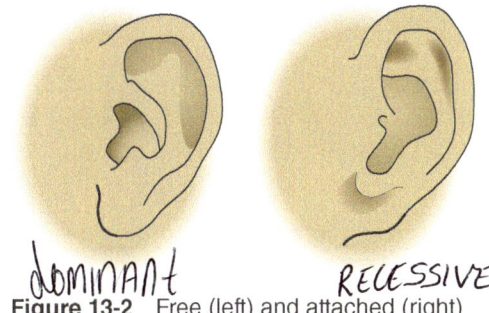

dominAnt   *RECESSIVE*

Figure 13-2 Free (left) and attached (right) earlobes in humans.

When both alleles in a nucleus are identical, the nucleus is **homozygous**. Those with both dominant alleles are homozygous dominant.

When both recessives are present in the same nucleus, the individual is said to be *homozygous recessive* for the trait.

When both the dominant and recessive alleles are present in a single nucleus, the individual is **heterozygous** for that trait.

A man has the genotype FF. What is the genotype of his gamete (sperm) nuclei? ___F___

A woman has attached earlobes. What is her genotype? ___ff___

What allele(s) does(do) her gametes (ova) carry? ___f___

These two individuals produce a child. Show the genotype of the child by diagramming the cross:

sperm genotype   ×   ovum genotype

___F___   ___f___

___Ff___
child's genotype   *FREE EARLOBES*

pheno - physical

geno - sequence

What is the phenotype of the child? (That is, does this child have attached or free earlobes?) ___FREE___

4. In garden peas, purple flowers are dominant over white flowers. Let A represent the allele for purple flowers, a the allele for white flowers.

(a) What is the phenotype (color) of the flowers with the following genotypes?

Genotype	Phenotype
AA	PURP
aa	WHITE
Aa	PURP

Note: Always distinguish clearly between upper- and lowercase letters.

A white-flowered garden pea is crossed with a homozygous dominant purple-flowered plant.

(b) Name the genotype(s) of the gametes of the white-flowered plant. ___aa___

(c) Name the genotype(s) of the gametes of the purple-flowered plant. ___AA___

(d) Name the genotype(s) of the plants produced by the cross. ___Aa___

(e) Name the phenotype(s) of the plants produced by the cross. ___PURP FLOWERS___

(f) The Punnett square is a convenient way to perform the mechanics of a cross. The circles along the top and side of the Punnett square represent the possible gamete nuclei. Insert the proper letters indicating the genotypes of the possible gamete nuclei for the white/purple flowered cross in the circles, then fill in the following Punnett square for all the possible genetic outcomes represented by each combination of gametes.

gametes of white-flowered plant

	(a)	(a)
(A)	Aa	Aa
(A)	Aa	Aa

gametes of purple-flowered plant

GENO: Aa
PHENO: PURP

(g) A heterozygous plant is crossed with a white-flowered plant. Fill in the Punnett square below, then give the genotypes and phenotypes of all the possible genetic outcomes.

gametes of white-flowered plant

	(a)	(a)
(A)	Aa	Aa
(a)	aa	aa

gametes of heterozygote

GENO: Aa, aa
  2:2 , 1:1
PHENO: PURP/WHITE
  50/50

Possible genotypes: ___Aa, aa___
Possible phenotypes: ___PURP OR WHITE___
Draw a line connecting each possible genotype listed above to its respective phenotype.

(h) It's unlikely that every cross between two pea plants will produce four seeds that will in turn grow into four offspring plants every time. Rather, one of the most useful facets of problems such as these is that they allow you to *predict* the chances of a particular genetic outcome occurring. Genetics is really a matter of **probability,** the likelihood of the occurrence of a particular outcome.

To take a simple example, consider that the probability of coming up with heads in a single toss of a coin is one chance in two, or ½. Now let's apply this idea to the probability that offspring will have a certain genotype. Look at your Punnett square in part (g). The probability of having a genotype is the sum of all occurrences of that genotype. For example, the genotype *Aa* occurs in two of the four boxes. The probability that the genotype *Aa* will be produced from that particular cross is thus ²⁄₄, or 50%.

(i) What is the probability of an individual from part (g) having the genotype *aa*? ___50 % , 1:1___

B. An Observable Monohybrid Cross (About 15 min.)

MATERIALS

Per student group (table):

- genetic corn ears illustrating monohybrid cross
- hand lens or magnifying lens (optional)

Examine the monohybrid genetic corn demonstration. This illustrates a monohybrid cross between plants producing purple kernels and ones producing yellow kernels. By convention, P stands for the parental generation. The offspring are called the *first filial generation*, abbreviated F_1. If these F_1 offspring are crossed, their offspring are the *second filial generation*, designated F_2, and are shown as follows:

PROCEDURE

$$P \times P \quad\quad \text{and} \quad\quad F_1 \times F_1$$
$$\downarrow \quad\quad\quad\quad\quad\quad\quad\quad \downarrow$$
$$F_1 \quad\quad\quad\quad\quad\quad\quad\quad F_2$$

1. Note that all the first-generation kernels (F_1) in the genetic corn demonstration are purple, while the second-generation ear (F_2) has both purple and yellow kernels. Count the purple kernels and then the yellow ones in the F_2 ear. _____ purple; _____ yellow. When reduced to the lowest common denominator, is this ratio closest to 1:1, 2:1, 3:1, or 4:1? _____ This is called the *phenotypic ratio*.

2. A corncob with kernels represents the products of multiple instances of sexual reproduction. Each kernel represents a single instance; fertilization of one egg by one sperm produced *each* kernel. Thus each kernel represents a different cross.

Let the letter P represent the gene for kernel color.

  (a) What genotypes produce a purple phenotype? _____

  (b) Which allele is dominant? _____

  (c) What is the genotype of the yellow kernels on the F_2 ear? _____

  (d) You are given an ear with purple kernels. How can you determine its genotype with a single cross? Explain.

  _____

  _____

  _____

C. Experiment: Monohybrid Heredity in a Fern

A significant limitation of carrying out genetics experiments in the biology lab is that most take several months or years to collect relevant data. Even though Mendel could raise two generations of peas in a growing season, the experiments conducted in his garden plot often lasted several years. Two growing seasons were required to produce the corn monohybrid cross studied above. Fortunately, we can now look at inheritance in organisms with a much shorter life cycle. We'll use C-ferns to investigate a monohybrid cross.

Like all ferns, C-ferns have two independent life cycle phases: a structurally simple, haploid gametophyte and a more complex diploid sporophyte. (See Figures 12-4 and 12-5, to review the generalized fern life cycle.) A mature C-fern plant produces haploid spores via the process of meiosis. The spores germinate under suitable environmental conditions, and begin to divide mitotically to produce the gametophyte phase (Figure 13-3). This haploid phase develops very rapidly, with gametophytes maturing within 2 weeks.

At maturity, the gametophyte consists of a small (2 mm), simple, photosynthetic flattened structure with sex organs that produce *by mitosis* eggs in structures called archegonia and/or sperm in structures called antheridia. In the presence of water, flagellated sperm are discharged. The sperm are attracted to substances produced by the archegonia and swim toward the egg. Eventually one sperm fertilizes the egg, producing the first cell of the next diploid sporophyte generation, the zygote.

The photosynthetic sporophyte also develops rapidly, with roots and leaves visible within 1–2 weeks. The C-fern sporophytes reach heights of 10–40+ cm. Spores are produced by meiosis in structures on the leaves, completing the life cycle.

C.1. Week 1—Observation of F_1 Hybrid Gametophytes (About 30 min.)

MATERIALS

Per student:

- 2-week-old F_1 C-fern gametophyte culture in petri dish
- dissecting microscope
- sterile dH_2O
- sterile pipet
- marking pen
- calculator (optional)

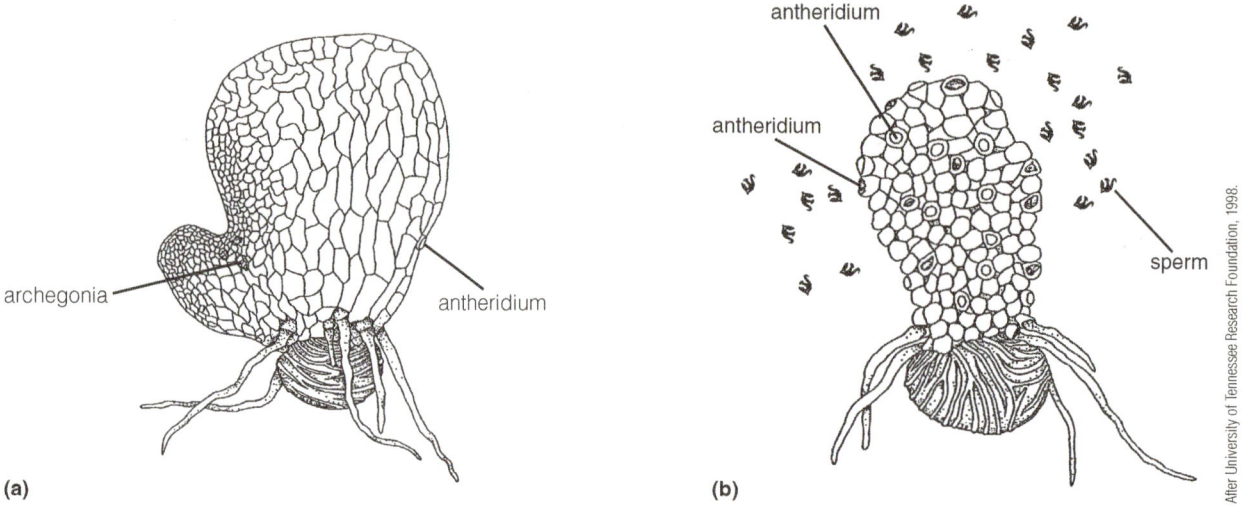

archegonia

antheridium

(a)

antheridium

antheridium

sperm

(b)

After University of Tennessee Research Foundation, 1998.

Figure 13-3 Mature C-fern gametophytes and gametes. (**a**) Hermaphroditic (bisexual) gametophytes produce both eggs and sperm and are somewhat heart-shaped. (**b**) Male gametophytes produce only sperm and appear tongue-shaped.

Prior to this class, petri dishes with nutrient medium were inoculated with spores from an F_1 hybrid C-fern sporophyte plant. The spores germinated and have grown into mature haploid gametophytes.

PROCEDURE

1. Observe the F_1 cultures under the dissecting microscope with the highest magnification possible and transmitted light (light from below.) You can prevent your cultures from drying out from the heat of the microscope by leaving the lid on the culture as much as possible and by turning the light off or removing the culture to the lab bench when it is not being observed.
2. While you are observing the culture, tilt the lid up and use a sterile pipet to add 1–2 mL sterile distilled water. Lower the lid and tilt the plate back and forth to cover all of the gametophytes with water. Observe the release of swimming sperm from antheridia and their attempts to find and fertilize mature eggs within archegonia; the sperm often spin in place briefly once they're released from an antheridium before zooming off and homing in on chemical signals produced by a mature archegonium.

   Do all the gametophytes have the same phenotype? Describe any differences you observe.

   _____

   _____

   _____

   For this experiment, we will focus only on the larger, heart-shaped hermaphroditic gametophytes. Which of the phenotypes would you designate as a mutant? Why?

   _____

   _____

3. Take a random sample of the hermaphroditic F_1 gametophyte population by counting up to 50 individuals and tallying their phenotypes. First, think of the following questions about collecting such data:
   (**a**) Why is it important to take a random sample from the cultures?

   _____

   (**b**) What is a suitable method of collecting data that would ensure a random sample?

   _____

   _____

4. Record your data in Table 13-1 and in the location designated by your instructor for the class data. Leave the lid on during this procedure if possible. If it becomes fogged, quickly exchange the fogged lid for a clean lid from an unused dish. After scoring, replace the old lid over the culture.
5. When you've finished your observations, remove any excess water from the culture by lifting the lid slightly and pouring off the excess. Place the culture in the location designated by your instructor. Be sure the petri dish lid is in place.
   (**a**) Are the F_1 plants in the culture dish haploid or diploid?

   _____

   (**b**) What products will result from the fertilization events in the culture?

   _____

TABLE 13-1 Gametophyte Phenotypes

Description of Phenotypes	Number of Gametophytes	Class Total

© Cengage Learning 2013

Will the products of fertilization be haploid or diploid?

_____

(c) If an F_1 sporophyte is heterozygous for a single mutant trait, use your own gene and allele designations to list the genotypes that will be present in the spores produced by that sporophyte.

_____

What genotypes will be present in the F_3 gametophytes that grow from those spores?

_____

(d) What is the *expected* ratio of genotypes in the F_3 gametophytes?

_____

(e) What is the approximate *actual* phenotypic ratio of the gametophytes you counted?

_____

(f) Can you determine the dominance relationships from the data in Table 13-1? Explain and describe.

_____

_____

Biologists have assigned the designation CP to the single gene responsible for the two different phenotypes seen in this experiment. The dominant allele is thus designated CP, and the recessive allele cp.

(g) Predict the genetic outcome of the fertilizations taking place in your culture by formulating a hypothesis to explain the inheritance of the trait. Indicate expected ratios of both the gametophyte and the F_2 generations.

_____

_____

C.2. Data Analysis (About 15 min.)

Gametes from true-breeding gametophyte parents (the P generation) combine to produce hybrid F_1 sporophyte fern plants. Meiosis within the F_1 plants produces spores. Each gametophyte you observe resulted from mitotic divisions of one of those spores. Their gametes will combine to produce the F_2 generation sporophytes.

In the space below, diagram the crosses involved in the F_1 and F_2 generations, indicating which generations and organisms are haploid, and which are diploid. (Refer to the crosses diagrammed in Part 13.1.B. as an example.)

You can now test your hypothesis concerning the method of trait inheritance to determine whether the data you collected support or do not support your model. Geneticists typically use the chi-square (χ^2) statistical test to determine whether experimentally obtained data are a satisfactory approximation of the expected data. In short, this test expresses the difference between expected (hypothetical) and observed (collected) numbers as a single value, χ^2. If the difference between observed and expected results is large, a large χ^2 results, while a small difference results in a small χ^2. Chi-square values are calculated according to the formula

$$\chi^2 = \sum \frac{(O - E)^2}{E}$$

where O = *observed* number of individuals
E = *expected* number of individuals
Σ = the sum of all values of $(O - E)^2/E$ for the various categories of phenotypes

Let's use this formula in our garden-pea monohybrid cross, question 4 from pages 163–164. Suppose 81 flowers are counted in a cross. Our hypothesis (expectation) is that three-fourths of them will be purple:

$$\frac{3}{4} \times 81 = 60.75$$

Similarly, we expect one-fourth to be white:

$$\frac{1}{4} \times 81 = 20.25$$

Suppose we actually count 64 purple flowers and 17 white flowers. Examine Table 13-2, noting how these values are used.

$$\chi^2 = \sum (0.174 + 0.522) = 0.696$$

TABLE 13-2	Calculations of Chi Square for Garden-Pea Monohybrid Cross					
Phenotype	Genotype	O	E	$(O - E)$	$(O - E)^2$	$(O - E)^2/E$
Purple	$P\_$	64	60.75	3.25	10.56	0.174
White	pp	17	20.25	−3.25	10.56	0.522
Total		81	81	0		0.696

© Cengage Learning 2013

Now, how do we interpret the χ^2 value we found? Suppose the expected and observed values were identical. Then $\chi^2 = 0$. You might guess that a number very close to zero indicates close agreement between observed and expected and a large χ^2 value suggests that "something unusual" is taking place. The problem is that chance alone almost always causes small deviations between observed and expected results, *even when the hypothesis being tested is correct.*

When does the χ^2 value indicate that chance alone cannot explain the deviation? Geneticists generally agree on a probability value of 1 in 20 (or 5% = .05) as the lowest acceptable value derived from the χ^2 test. This number indicates that if the experiment is repeated many times, the deviations expected due to chance alone will be as large as or larger than those observed only about 5% or less of the time. Probabilities equal to or greater than .05 are considered to support the hypothesis, while probabilities lower than .05 do **not** support the hypothesis. Here we must consult a table of χ^2 values to make our decision (Table 13-3).

In our example, the χ^2 value is 0.696. Since this is a monohybrid problem with only two categories of possible outcomes (purple or white flowers), the number of degrees of freedom (*n* in the left-hand column of Table 13-3) is 1. Read across the table until you come to .05 and find the χ^2 value 3.84. Because 0.696, our calculated χ^2 value, is less than 3.841, it is likely that the variation in the observed and expected is the result of chance, and that our hypothesized outcome is correct. A value *greater than* 3.841, however, would indicate that chance alone cannot explain the deviation between observed and expected, and we would reject our hypothesis.

The term *degrees of freedom* requires further explanation. The number of degrees of freedom is always 1 *less* than the number of categories of possible outcomes. Thus, if you are dealing with a dihybrid problem with a ratio of 9:3:3:1 (four possible phenotypes), *n* =3.

TABLE 13-3	Distribution of χ^2			
	Probability of Obtaining a χ^2 Value as Large or Larger			
Degrees of Freedom, *n*	.10	.05	.01	.001
1	2.71	3.84	6.63	10.83
2	4.61	5.99	9.21	13.82
3	6.25	7.82	11.35	16.27
4	7.78	9.49	13.28	18.47

© Cengage Learning 2013

1. Transfer your individual or group data from the Total columns in Table 13-1 to Table 13-4 and calculate χ^2.

TABLE 13-4	χ^2 Calculation from F_1 Gametophyte Data				
Phenotype	Observed (O)	Expected (E)[a]	(O − E)	(O − E)^2	(O − E)^2/E
Totals					$\chi^2 =$

© Cengage Learning 2013

[a]This number should be based on the hypothesis you developed in week 1 observations pages 166–167.

2. Use Table 13-3 to determine the probability of obtaining this χ^2 value for the gametophyte data in Table 13-4. How many degrees of freedom are there? _____

Is your hypothesis supported or not supported? _____ If not, what might be changed in your hypothesis or in the experimental design?

_____

_____

C.3. Week 3—Observation of F_2 Sporophytes (About 45 min.)

MATERIALS

Per student:

- 4-week-old F_2 C-fern sporophyte culture in petri dish
- dissecting microscope
- dissecting needle or toothpicks
- calculator (optional)

PROCEDURE

1. Examine your cultures with the dissecting microscope. Mutant and wild-type phenotypes are best observed using reflected light from the top or the side. Carefully observe the oldest leaves. Can you see mutant and wild-type phenotypes? _____
   Are the young sporophytes haploid or diploid? How do you know?

_____

_____

2. Sketch what you are observing, and label it with the following terms: gametophyte, sporophyte leaf, sporophyte root.
3. Take a random sample of the sporophyte population in a dish by counting up to 50 individuals and identifying their phenotype. You can remove the lid from the culture to do this. It may be easier to score the phenotype after gently and randomly pulling up individual sporophytes with a dissecting needle or toothpick and laying them out in a row on empty areas of the culture plate. Observe the largest leaf on each sporophyte and examine the differences carefully before recording data in Table 13-5 and in the location designated for class data.
4. Following scoring of phenotypes, place the lid back on the plate and return the culture to the designated location, or take it home so that you can observe it over the next several weeks to determine whether the phenotype of older sporophytes is apparent without use of a microscope.

TABLE 13-5	F$_2$ Sporophyte Phenotypes	
Description of Phenotypes	**Number of Sporophytes**	**Class Total**

© Cengage Learning 2013

5. Restate your hypothesis regarding the inheritance of the mutant and wild-type alleles, and your prediction of the genetic outcome in the F$_2$ sporophytes.

_____

_____

_____

6. Transfer your individual or class total data to Table 13-6 to calculate χ^2.

TABLE 13-6	χ^2 Calculation from F$_2$ Sporophyte Data				
Phenotype	**Observed (O)**	**Expected (E)^a**	**(O – E)**	**(O – E)^2**	**(O – E)^2/E**
Totals					$\chi^2=$

© Cengage Learning 2013

_____

^{a}This number should be based on the hypothesis you developed in this week's observations.

7. Use Table 13-3 to determine the probability of obtaining this χ^2 value for the sporophyte data in Table 13-6.
   How many degrees of freedom are there? _____
   What is the approximate probability? _____
   Is your hypothesis supported or not supported? _____
   Which allele is the dominant allele? _____
   Which is recessive? _____
   If gametophytes had not expressed the phenotype, would you be able to form a hypothesis from observations of the gametophyte generation? _____ Why or why not? _____

   _____

13.2   Dihybrid Inheritance

All the problems and experiments so far have involved the inheritance of only one trait; that is, they are monohybrid problems. Now we'll examine cases in which two traits are involved: **dihybrid problems.**

Note: **We will assume that the genes for these traits are carried on different (nonhomologous) chromosomes.**

A. Dihybrid Problems (About 15 min.)

1. Let's consider these two traits:
   - In humans, a pigment in the front part of the eye masks a blue layer at the back of the iris. The dominant allele _P_ causes production of this pigment. Those who are homozygous recessive _(pp)_ lack the pigment,

and the back of the iris shows through, resulting in blue eyes. (Other genes determine the color of the pigment, but in this problem we'll consider only the presence or absence of *any* pigment at the front of the eye.)

- Dimpled chins (*D* = allele for dimpling) are dominant over undimpled chins (*d* = allele for lack of dimple).

    (a) List all possible genotypes for an individual with pigmented iris and dimpled chin. *4 PPDD*

    (b) List the possible genotypes for an individual with pigmented iris but lacking a dimpled chin. *2 PPdd   Ppdd*

    (c) List the possible genotypes of a blue-eyed, dimple-chinned individual. *Pp Dd   ppDD*

    (d) List the possible genotypes of a blue-eyed individual lacking a dimpled chin. *Ppdd, Pd, ppdd*

2. An individual is heterozygous for both traits (eye pigmentation and chin form).

    (a) What is the genotype of such an individual? *PpDd*

    (b) What are the possible genotypes of that individual's gametes? *Pd, PD, pd, pD*

        If determining the answer for question 2 was difficult, recall from Exercise 12 that the principle of independent assortment states that genes on different (nonhomologous) chromosomes are separated out independently of one another during meiosis. That is, the occurrence of an allele for eye pigmentation in a gamete has *no bearing* on which allele for chin form will occur in that same gamete.

        There is a useful method for determining possible gamete genotypes produced during meiosis from a given parental genotype. Using the genotype *PpDd* as an example, follow the four arrows below to determine the four possible gamete genotypes:

Pp Dd, PPDd, PPDD

$$PpDd$$

    (c) Two individuals heterozygous for both eye pigmentation and chin form have children. What are the possible genotypes of those F_1 offspring?

PpDd, PPDd, ppDd, ppdd

        You can set up a Punnett square to do dihybrid problems just as you did with monohybrid problems. However, depending on the parental genotypes, the square may have as many as 16 boxes, rather than just 4. Insert the possible genotypes of the gametes from one parent in the top circles and the gamete genotypes of the other parent in the circles to the left of the box.

gametes of one parent

	PD	Pd	pD	pd
PD	PPDD	PPDd	PpDD	PpDd
Pd	PPDd	PPdd	PpDd	Ppdd
pD	PpDD	PpDd	PpDD	ppDd
pd	PpDd	Ppdd	ppDd	ppdd

gametes of other parent

    (d) Possible genotypes of children produced by two parents heterozygous for both eye pigmentation and chin form: *4*

        What is the ratio of the genotypes? *1:2:2:4:1:2:1:2:1*
        What is the phenotypic ratio? *9 : 3 : 3 : 1   Blue/DIMPLE*

    (e) Recalling the discussion of probability in Section 13.1, state the probability of a child from part **(d)** having the following genotypes.

        ppDD   *1/16*

        PpDd   *5/16*

        PPDd   *1/16*

        To extend the probability discussion, let's reconsider flipping a coin by asking the question, What is the probability of flipping heads twice in a row? The chance of flipping heads the first time is ½. The same is true for the second flip. The chance (probability) that we'll flip heads twice in a row is ½ x ½ = ¼. The probability that we could flip heads three times in a row is ½ x ½ x ½ = ⅛.

    (f) State the probability that three children born to the parents in part **(d)** will have the genotype *ppdd*.

        1/16

        What is the probability that three children born to these parents will have dimpled chins and pigmented eyes? *9/16*

    (g) What is the genotype of the F_1 generation when the father is homozygous for both pigmented eyes and dimpled chin, but the mother has blue eyes and no dimple? *PPDD , Ppdd*

        What is the phenotype of this individual? *BLUE EYES , NO DIMPLE*

B. An Observable Dihybrid Cross (About 20 min.)

MATERIALS

Per student group (table):

- genetic corn ears illustrating a dihybrid cross

PROCEDURE

1. Examine the demonstration of dihybrid inheritance in corn. Notice that not only are the kernels two different colors (one trait), but they are also differently shaped (second trait). Kernels with starchy endosperm (the carbohydrate-storing tissue) are smooth, while those with sweet (sugary) endosperm are shriveled. Notice that all *four* possible phenotypic combinations of color and shape are present in the F_2 generation.

   The P gene is involved in pigment production, with two alleles P and p. The S gene determines carbohydrate (sugar) storage, with two alleles S and s.

   Which genotypes of the parents produced the F_2 generation kernels? _____

2. Set up a Punnett square of this dihybrid cross:

   What is the predicted phenotypic ratio? _____

3. Count the number of kernels of each possible phenotype and record in Table 13-7. To increase your sample size, count three ears.

   Which traits seem dominant?

   _____

   Which traits seem recessive?

   _____

4. Calculate the actual phenotypic ratio you observed:

   _____

   Do your observed results differ from the expected results? _____

gametes of one parent

gametes of other parent

TABLE 13-7	Phenotypes in Dihybrid Corn Cross			
	Number of Kernels with Phenotypes			
Ear	**Yellow Smooth**	**Yellow Shriveled**	**Purple Smooth**	**Purple Shriveled**
1				
2				
3				
Totals				

© Cengage Learning 2013

5. Use the chi-square test to determine if the deviation from the expected results can be accounted for by chance alone.

   Chi-square test results: _____

   _____

13.3   Some Readily Observable Human Traits *(About 15 min.)*

In the preceding pages, we examined several human traits that are fairly simple and that follow the Mendelian pattern of inheritance. Most of our traits are much more complex, involving many genes or interactions between genes. For example, hair color is determined by at least four genes, each one coding for the production of melanin, a brown pigment. Because the effect of these genes is cumulative, hair color can range from blond (little melanin) to very dark brown (much melanin).

Clearly, human traits are of great interest to us. Table 13-8 lists a number of traits that seem to exhibit Mendelian inheritance. For each trait, work with a lab partner to determine your phenotype, then record in Table 13-8. List your possible genotype(s) for each trait. When convenient, examine your parents' phenotypes and attempt to determine your actual genotype.

TABLE 13-8	Summary of My Mendelian Traits							
			Mom's		Dad's			
Trait	My Phenotype	My Possible Genotypes	Phenotype	Possible Genotype	Phenotype	Possible Genotype	My Possible/ Probable Genotype
Mid-digital hair							
Tongue rolling							
Widow's peak							
Earlobe attachment							
Hitchhiker's thumb							
Relative finger length							

© Cengage Learning 2013

1. *Mid-digital hair* (Figure 13-4a). Examine the joint of your fingers for the presence of hair, the dominant condition (*MM*, *Mm*). Complete absence of hair is due to the homozygous-recessive condition (*mm*). You may need a hand lens to determine your phenotype. Even the slightest amount of hair indicates the dominant phenotype.

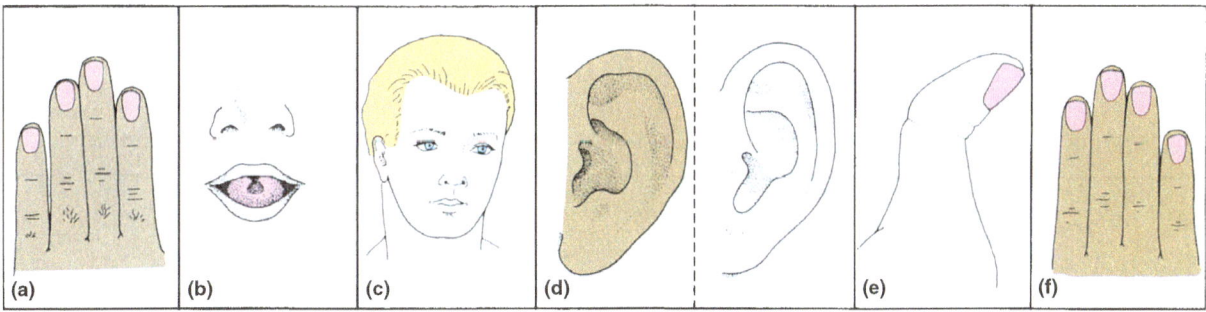

Figure 13-4   Some readily observable human Mendelian traits.

2. *Tongue rolling* (Figure 13-4b). The ability to roll one's tongue is due to a dominant allele, *T*. The homozygous-recessive condition, *tt*, results in inability to roll the tongue (Figure 13-5).
3. *Widow's peak* (Figure 13-4c). Widow's peak describes a distinct downward point in the frontal hairline and is due to the dominant allele, *W*. The recessive allele, *w*, results in a continuous hairline. (Omit study of this trait if baldness is affecting the hairline.)
4. *Earlobe attachment* (Figure 13-4d). Most individuals have free (unattached) earlobes (*FF*, *Ff*). Homozygous recessives (*ff*) have earlobes attached directly to the head.
5. *Hitchhiker's thumb* (Figure 13-4e). Although considerable variation exists in this trait, we'll consider those

Graham Dunn/Alamy

Figure 13-5   A family with a tongue-rolling father and tongue-rolling-challenged mother. What are the genotype and phenotype of their son?

individuals who *cannot* extend their thumbs backward to approximately 45° to be carrying the dominant allele, *H*. Homozygous-recessive persons (*hh*) can bend their thumbs at least 45°, if not farther.

6. *Relative finger length* (Figure 13-4f). An interesting sex-influenced (*not* sex-linked) trait relates to the relative lengths of the index and ring finger. In males, the allele for a short index finger (*S*) is dominant. In females, it's recessive. In rare cases, each hand is different. If one or both index fingers are greater than or equal to the length of the ring finger, the recessive genotype is present in males, and the dominant one present in females.

Genetics Problems. One of the best ways to solidify your understanding of different patterns of inheritance is to work genetics problems. Your instructor may assign you portions of Appendix 2, which is a collection of these useful and interesting problems.

_A_ 1. In a monohybrid cross,
(a) only one trait is being considered.
(b) the parents are always dominant.
(c) the parents are always heterozygous.
(d) no hybrid is produced.

_B_ 2. The genetic makeup of an organism is its
(a) phenotype.
(b) genotype.
(c) locus.
(d) gamete.

_D_ 3. An allele whose expression is completely masked by the expression or effect of its allelic partner is
(a) homologous.
(b) homozygous.
(c) dominant.
(d) recessive.

_A_ 4. The physical appearance and physiology of an organism, resulting from interactions of its genetic makeup and its environment, is its
(a) phenotype.
(b) hybrid vigor.
(c) dominance.
(d) genotype.

_D_ 5. When both dominant and recessive alleles are present within a single nucleus, the organism is _____ for the trait.
(a) diploid
(b) haploid
(c) homozygous
(d) heterozygous

_D_ 6. A Punnett square is used to determine
(a) probable gamete genotypes.
(b) possible parental phenotypes.
(c) possible parental genotypes.
(d) possible genetic outcomes of a cross.

_D_ 7. The gametophyte of a fern is
(a) haploid.
(b) photoautotrophic.
(c) a structure that produces eggs and/or sperm.
(d) all of the above

_A_ 8. A chi-square test is used to
(a) determine if experimental data adequately matches what was expected.
(b) analyze a Punnett square.
(c) determine parental genotypes producing a given offspring genotype.
(d) determine if a trait is dominant or recessive.

_D_ 9. Possible gamete genotypes produced by an individual of genotype PpDd are
(a) Pp and Dd.
(b) all PpDd.
(c) PD and pd.
(d) PD, Pd, pD, and pd.

_A_ 10. If you can roll your tongue,
(a) you have at least one copy of the dominant allele T.
(b) you have two copies of the recessive allele t.
(c) you must be male.
(d) you are haploid.

EXERCISE **1 3**

Heredity

Post-Lab Questions

13.1 Monohybrid Crosses

1. Explain how Mendel's law of segregation applies to the distribution of alleles in gametes.

2. Assume that production of hairs on a plant's leaves is controlled by a single gene with two alleles, H (dominant) and h (recessive). Hairy leaves are dominant to smooth (nonhairy) leaves.
    (a) Name the genotype(s) of a smooth-leaved plant. _____
    (b) Name the genotype(s) of a hairy-leaved plant. _____
    (c) What are the possible genotypes of gametes produced by the smooth-leaved plant? _____
    (d) What are the possible genotypes of gametes produced by the hairy-leaved plant? _____

3. *Non*-true-breeding hairy-leaved plants are crossed with smooth-leaved plants.
    (a) What genotypic and phenotypic ratios would you expect for the potential offspring? _____
    (b) Suppose you perform such a cross, collect data, and do a chi-square test to aid in data analysis. How many degrees of freedom would there be? _____
    (c) Suppose your chi-square value is very large (>25). What does this indicate about your experiment and/or hypothesis?

4. What genotypic ratio would you expect in the gametophyte generation of C-ferns produced by F_1 spores if two traits on separate chromosomes were being followed?

5. Were dominant and recessive traits observed equally in both gametophytes and sporophytes of C-ferns? How did you determine which character was dominant and which was recessive?

13.2 Dihybrid Inheritance

6. Suppose you have two traits controlled by genes on separate chromosomes. If sexual reproduction occurs between two heterozygous parents, what is the genotypic ratio of all possible gametes?

Food for Thought

7. Explain the purpose and uses of the chi-square test.

8. Suppose students in previous semesters had removed some of the corn kernels from the genetic corn ears before you counted them. What effect would this have on your results?

9. Assume that one allele is completely dominant over the other for the following questions.
   (a) Two individuals heterozygous for a *single* trait have children. What is the expected phenotypic ratio of the possible offspring? _____
   (b) Two individuals heterozygous for *two* traits have children. What would be the expected phenotypic ratio of the possible offspring? _____
   (c) Crossing two individuals heterozygous for two traits results in the same phenotypic ratio as for a single trait. Are the genes for these two traits on separate chromosomes or on the same chromosome? Explain your answer. (Remember that the gene for each trait is located at a locus, a physical region on the chromosome.)

10. How does probability differ from actuality?

Measurement Conversions

Metric to American Standard	*American Standard to Metric*
Length	*Length*
1 mm = 0.039 inch	1 inch = 2.54 cm
1 cm = 0.394 inch	1 foot = 0.305 m
1 m = 3.28 feet	1 yard = 0.914 m
1 m = 1.09 yards	1 mile = 1.61 km
1 km = 0.622 miles	
Volume	*Volume*
1 mL = 0.0338 fluid ounce	1 fluid ounce = 29.6 mL
1 L = 4.23 cups	1 cup = 237 mL
1 L = 2.11 pints	1 pint = 0.474 L
1 L = 1.06 quarts	1 quart = 0.947 L
1 L = 0.264 gallon	1 gallon = 3.79 L
Mass	*Mass*
1 mg = 0.0000353 ounce	1 ounce = 28.3 g
1 g = 0.0353 ounce	1 pound = 0.454 kg
1 kg = 2.21 pounds	

The Scientific Method

To appreciate biology or, for that matter, any body of scientific knowledge, you need to understand how the **scientific method** is used to gather that knowledge. We use the scientific method to test the predictions of possible answers to questions about nature in ways that we can duplicate or verify. Answers supported by test results are added to the body of scientific knowledge and contribute to the concepts presented in your textbook and other science books. Although these concepts are as up-to-date as possible, they are always open to further questions and modifications.

One of the roots of the scientific method can be found in ancient Greek philosophy. The natural philosophy of Aristotle and his colleagues was mechanistic rather than vitalistic. A **mechanist** believes that only natural forces govern living things, along with the rest of the universe, while a **vitalist** believes that the universe is at least partially governed by supernatural powers. Mechanists look for interrelationships between the structures and functions of living things, and the processes that shape them. Their explanations of nature deal in **cause and effect**—the idea that one thing is the result of another thing (for example, fertilization of an egg initiates the developmental process that forms an adult.). In contrast, vitalists often use purposeful explanations of natural events (the fertilized egg strives to develop into an adult). Although statements that ascribe purpose to things often feel comfortable to the writer, try to avoid them when writing lab reports and scientific papers.

Aristotle and his colleagues developed three rules to examine the laws of nature: Carefully observe some aspect of nature; examine these observations as to their similarities and differences; and produce a principle or generalization about the aspect of nature being studied (for example, all mammals nourish their young with milk).

The major defect of natural philosophy was that it accepted the idea of *absolute truth*. This belief suppressed the testing of principles once they had been formulated. Thus, Aristotle's belief in spontaneous generation, the principle that some life can arise from nonliving things (say maggots from spoiled meat), survived over 2000 years of controversy before being discredited by Louis Pasteur in 1860. Rejection of the idea of absolute truth coupled with the testing of principles either by experimentation or by further pertinent observation is the essence of the modern scientific method.

Although there is not one universal scientific method, Figure A-1 illustrates the general process.

MATERIALS

Per lab room:

- blindfold
- plastic beakers with an inside diameter of about 8 cm stuffed with cotton wool
- four or five liquid crystal thermometers

Step 1. Observation. As with natural philosophy, the scientific method starts with careful observation. An investigator may make observations from nature or from the written words of other investigators, which are published in books or research articles in scientific journals and are available in the storehouse of human knowledge, the library. One subject we all have some knowledge of is the human body. The first four rows of Table 1 list some observations about the human body. The fifth row is blank so that you can fill in the steps of the scientific method for some observation about the human body, or anything else you and your instructor wish to investigate.

Step 2. Question. In the second step of the scientific method, *we ask a question* about these observations. The quality of this question will depend on how carefully the observations were made and analyzed. Table 1 includes questions raised by the listed observations.

Step 3. Hypothesis. Now we *construct a hypothesis*—that is, we derive by inductive reasoning a possible answer to the question. **Induction** is a logical process by which all known observations are combined and considered before producing a possible answer. Table 1 includes examples of hypotheses.

Step 4. Prediction. In this step we *formulate a prediction*—we assume the hypothesis is correct and predict the result of a test that reveals some aspect of it. This is deductive or "if-then" reasoning. **Deduction** is a logical process by which a prediction is produced from a possible answer to the question asked. Table 1 lists a prediction for each hypothesis.

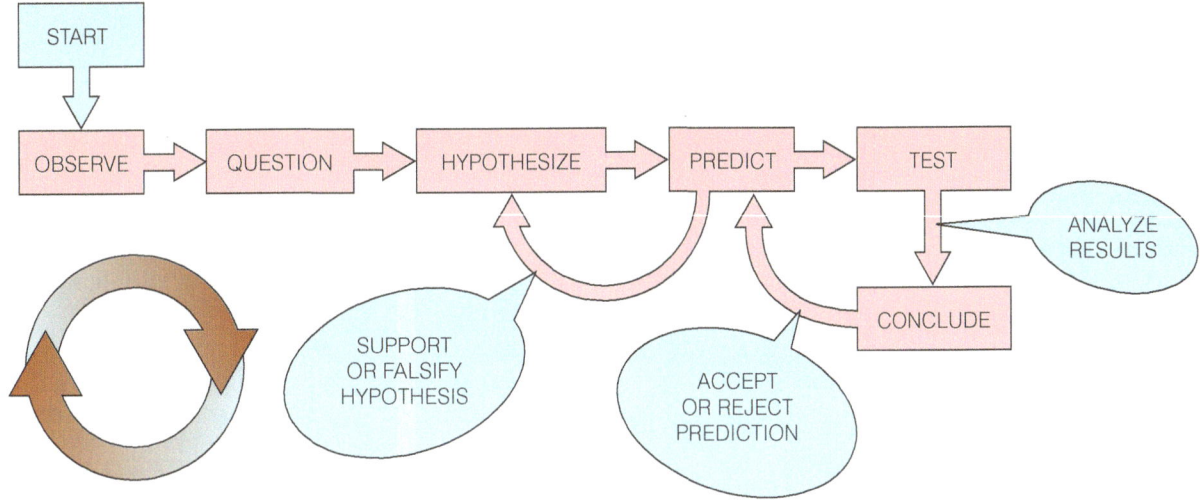

Figure A-1 The scientific method. Support or falsification of the hypothesis usually necessitates further observations, adjustment to the question, and modification of the hypothesis. Once started, the scientific method cycles over and over again, each turn further refining the hypothesis.

Step 5. Experiment or Pertinent Observations. In this step we *perform an experiment or make pertinent observations* to test the prediction. In an experiment of classical design, the individuals or items under study are divided into two groups: an **experimental group** that is treated with (or possesses) the independent variable and a **control group** that is not (or does not). Sometimes there is more than one experimental group. Sometimes subjects participate in both groups, experimental and control, and are tested both with and without the treatment.

In any test there are three kinds of variables. The **independent variable** is the treatment or condition under study. The **dependent variable** is the event or condition that is measured or observed when the results are gathered. The **controlled variables** are all other factors, which the investigator attempts to keep the same for all groups under study.

***Note:* The italicized statements show how the scientific method is applied to the predictions in Table 1 or give examples of the scientific method in practice.**

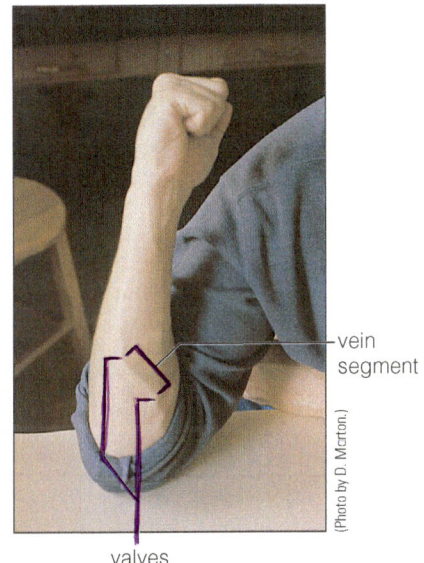

Figure A-2 Veins under the skin.

To test the prediction that blocking hearing in one ear impairs our ability to point out a sound's source (row II in Table 1), a group of subjects is tested first with no ears blocked (control group) and then with one ear blocked (experimental group). The independent variable is the blocking of one ear; the dependent variable is the ability to point out a sound's source; and the controlled variables are the standard conditions used for each trial—same test sound, same background noise, same procedure for each subject (same blocked ear, same instructions, same sequence of trials, same time between trials), and recruitment of appropriate subjects.

Sometimes the best tests of the predictions of a hypothesis are not actual experiments but further pertinent observations. One of the most important principles in biology, Darwin's theory of natural selection, was developed by this nonexperimental approach. Although they are a little more difficult to form, the nonexperimental approach also has variables.

To test the prediction that a liquid crystal thermometer will record different temperatures on the forehead, back of neck, and forearm (row IV of Table 1), the independent variable is location; the dependent variable is temperature; and the controlled variables are using the same thermometer to measure skin temperature at the three locations and measuring all of the subjects at rest.

Step 6. Conclusion. *To make a conclusion*—the last step in one cycle of the scientific method—you use the results of the experiment or pertinent observations to evaluate your hypothesis. If your prediction does not occur,

TABLE 1 Some Observations About the Human Body

Observation	Question	Hypothesis	Prediction
I. Veins containing blood are seen under the skin. Swellings present along the vein are often located where veins join together.[a]	What is the function of the swellings?	Swellings contain one-way valves that allow blood to flow only toward the heart.	If these valves are present, then blood flows only from vein segments farther from the heart to the next segments nearer the heart and never in the opposite direction.
II. People have two ears.	What is the advantage of having two ears?	Two ears allow us to locate the sources of sounds.	If the hypothesis is correct, then blocking hearing in one ear will impair our ability to determine a sound's source.[b]
III. People can hold their breath for only a short period of time.	What factor forces a person to take a breath?	The buildup of carbon dioxide derived from the body's metabolic activity stimulates us to take a breath.	If the hypothesis is correct, then people will hold their breath a shorter time just after exercise compared to when they are at rest.
IV. Normal body temperature is 98.6°F.	Is all of the body at the same 98.6°F temperature?	The skin, or at least some portion of it, is not 98.6°F.	If the hypothesis is correct, then a liquid crystal thermometer will record different temperatures on the forehead, back of neck, and forearm.
V. _____	_____	_____	_____

[a] The portion of the vein between swellings is called a segment and is illustrated in Figure A-2.
[b] This is especially the case for high-pitched sounds because higher frequencies travel less easily directly through tissues and bones.

it is rejected and your hypothesis or some aspect of it is falsified. If your prediction does occur, you may conditionally accept your prediction and your hypothesis is supported. However, you can never completely accept or reject any hypothesis; all you can do is state a probability that one is correct or incorrect. To quantify this probability, scientists use a branch of mathematics called _statistical analysis._

Even if the prediction is rejected, this does not necessarily mean that the treatment caused the result. A coincidence or the effect of some unforeseen and thus uncontrolled variable could be causing the result. For this reason, the results of experiments and observations must be _repeatable_ by the original investigator and others.

Even if the results are repeatable, this does not necessarily mean that the treatment caused the result. _Cause and effect,_ especially in biology, is rarely proven in experiments. We can, however, say that the treatment and result are correlated. A **correlation** is a relationship between the independent and the dependent variables.

Severe narrowing of a coronary artery branch reduces blood flow to the heart muscle downstream. This region of heart muscle gets insufficient oxygen and cannot contract and may die, resulting in a heart attack. The initial cause is the narrowing of the artery and the final effect is the heart attack. Perhaps the heart attack victim smoked cigarettes. Smoking cigarettes is one of several factors that make a person more likely to have a heart attack. This is based on a correlation between smoking and heart attacks in the general population but we cannot say for sure that the smoking caused the heart attack.

THEORIES and PRINCIPLES: When exhaustive experiments and observations consistently support an important hypothesis, it is accepted as a **theory**. A theory that stands the test of time may be elevated to the status of a **principle**. Theories and principles are always considered when new hypotheses are formulated. However, like hypotheses, theories and principles can be modified or even discarded in the light of new knowledge. Biology, like life itself, is not static but is constantly changing.

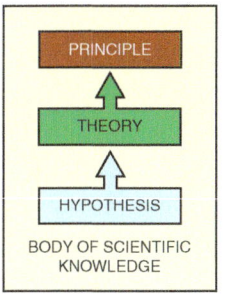

Scientific Writing and Critiquing

The account of one or several related cycles of the scientific method is usually reported in depth in a research article published in a scientific journal. Writing such research articles allows scientists to share knowledge. They provide enough information so that other scientists can repeat the experiments or pertinent observations they describe. The journal *Science,* along with several others, presents its research articles in narrative form, and many of the details of the scientific method are understood and not stated. However, adherence to the modern scientific method is expected, and the scientific community understands that it is as important to expose mistakes as it is to praise new knowledge.

Most scientific journals present research reports in a standard format. This allows scientists to quickly and easily skim through the many new research articles published daily, reading in depth only those of particular interest. By understanding the purpose of each section of a research article, you not only will be able to locate the question, research context, main methods, results, and meaning of someone's research with speed and efficiency, but also learn how to design and communicate your own research investigations along the way! After all, the best way to learn science is to *do* science—reading research articles, exploring and trying out various experimental and observational methodologies that you are reading about, and eventually building on and extending previous research that you have read that particularly interested you. Scientists rarely investigate topics out of the blue; most often they become involved in research that they have either read about or observed that interested them. They then browse the available scientific knowledge base, most often the scientific journal literature, to see what is already known about the topic of interest. If they come across an area where nothing is known—an area of ignorance—then by using the scientific method they may choose to extend our knowledge.

The following sections are most commonly used in the scientific literature. The word counts are suggestions only and offered as guidelines to use when writing your own research reports.

Abstract (~200 words): The abstract contains a brief and concise summary of the main sections of the scientific report. Usually included are the research questions, the broad conceptual framework or relevance of the research topic, the hypothesis and prediction(s), the main methods, the main results and their interpretation, and a concluding statement.

Introduction (~200–300 words): The introduction is a short narrative that describes the research topic in some detail. It explains what is known regarding the research topic, based on previous research. This section concludes with what is not known regarding the research topic, the specific question that the research report focuses on, the hypothesis and prediction(s), and possibly a brief description of how the question will be answered.

Materials and Methods (~200 words): This section explains how the research was conducted. It should provide enough detail that someone could read the research report and repeat the experiment. The statistical analyses that will be used are also described in this section.

Results (~150 words): This section reports the results of the research but does not go into detail explaining or discussing what they mean (that goes in the next section). Here is where any tables, graphs, or figures would be located. The narrative of this section explains what the tables, graphs, or figures show.

Discussion (~200–300 words): In this section, the research question is restated, and the answer that was obtained from the research is given. Data reported in the results section are clearly related to the hypothesis and predictions. The results are interpreted, any unexpected or inconsistent results are explained, and a discussion of what the results mean is provided. The results are also integrated with the work of others, relating the research conducted to the larger body of work already completed and reported in the scientific literature. Finally, the research is related to the big picture or larger conceptual framework within which the research topic lies. New research questions or avenues of research are mentioned as well.

Citations or References: The sources of information that were used in the research report are cited in the text by author, and the details for how to locate the information is given.

It is recommended that you read several research articles to get an idea of how research is reported. This will increase your ability to read and interpret scientific literature as well as increase your ability to communicate through writing. The following is a general guide for critiquing scientific manuscripts.

Evaluating a Manuscript

When evaluating a manuscript, it is important to consider several things. Some of the following questions can only be answered by those who are intimately familiar with the organisms and techniques used. Others concern the structure, style, and soundness of a manuscript and can be answered by anyone with a peripheral knowledge of the subject matter. The ability to consider any of these problems will, of course, increase with experience.

ABSTRACT

Is it informative, concise, and complete?

INTRODUCTION

What is the main point of this study?

Is its relevance to scientific theory (a conceptual framework) apparent?

If it contains a hypothesis, is it stated in a way that is testable?

METHODS

Can the work be repeated on the basis of the methods given?

Are unwieldy technical terms defined?

Are the methods and organisms used appropriate for testing the hypothesis?

To what extent can conclusions drawn from laboratory experiments be extended to interpret events occurring in nature?

RESULTS

Are all important results given?

Were the results properly evaluated statistically?

Are the figures and tables clear?

Do the results contained in the figures and tables support the statements made in the text?

DISCUSSION

Are the data clearly related to the hypothesis?

Are the results properly interpreted?

Are the results integrated with those of other workers?

Are unexpected or inconsistent results explained?

Does the paper provide any new ideas or interpretations?

If the discussion contains any speculation, is it justifiable?

The following is a sample checklist that can be used for peer reviews of scientific reports.

Checklist for Journal Article Peer Review

Clearly critique how well the author sets his or her research topic in context with other scientific studies (the conceptual framework of the article and research).
Use the following outline for critiquing the introduction:

Background overview

Conceptual framework

Specific focus of article (hypothesis and objectives, or premise of a review article)

Significance of specific focus

Synthesis of literature on specific topic as provided by the article's author(s)

Broader impact and relevance of the research question

Evaluate the following:

a. Introduction

_____ Does the author of the article provide a sufficient context or background for the study, relating the specific focus to previous research or conceptual development?

_____ Does the author link his or her research topic to some overarching theory or conceptual framework?

_____ Are the hypothesis and objectives (the specific focus) clear?

_____ Is there a clear rationale and justification for the study (i.e., what is the significance of the study topic, in terms of advancing knowledge and understanding within its own field or across different fields of science)?

b. Methods, results, and discussion

_____ Critique the article (not just summarize), discussing such things as the development of ideas pertaining to the study topic, criteria for evidence, appropriateness and strength of research methodologies, and so on.

c. Conclusions

_____ Discuss the main conclusions (take-home message) of the article.

d. Broader impact

_____ Does the author clearly discuss how the article advances discovery and understanding within or across fields of scientific study (does the article move back to the big picture beyond the specific focus and back to the conceptual framework)?

_____ Does the author provide examples of application and use for the study's ideas, concepts, and findings?

Biological question(s) you came up with from reading the article and its conclusions—that is, questions that may lead to future research:

_____
_____
_____
_____
_____
_____
_____
_____
_____
_____
_____
_____
_____
_____

APPENDIX 4

Statistics and Graphing with Excel

The following are references for commonly used statistical tests.

Nonparametric Tests

Nonparametric statistical tests do not require certain mathematical distributions of the data being analyzed. These tests can be used for measurement or count data as well as data that do not show a normal distribution.

Goodness-of-Fit Test (Chi-Square)

See discussion of chi-square in Exercise 9, section B.2.

Contingency Analysis (Cross-Tabulation of Frequency)

Contingency analysis is used with count data (frequencies) that are arranged in two or more rows, such as recording individuals by their species and their habitat location. This test is used to investigate the association between variables.

The Binomial Test

The binomial test is used when each individual in the sample is classified in one of two categories and the investigator wants to know if the proportion of individuals falling in each category differs from chance or from some previously specified probabilities of falling into those categories.

Parametric Tests

Parametric tests are based on certain assumptions of mathematical distribution of the data being analyzed. The most commonly used distribution is the normal distribution (other tests involve the binomial or Poisson distribution).

Comparison of Variances and Means for Two Samples: The t-Test

The *t*-test is one of the most widely used test procedures for comparing two samples to investigate whether significant differences exist between the two populations sampled.

Using Excel

Excel is an easy-to-use spreadsheet for creating data tables, calculating descriptive statistics, and graphing data. There are many useful online references that provide tutorials and walk you through using Excel. The following are recommended:

The Excel home page: http://office.microsoft.com/en-us/excel/default.aspx

An Excel tutorial: http://www.usd.edu/trio/tut/excel/

How to use Excel: http://serc.carleton.edu/introgeo/mathstatmodels/xlhowto.html

Using Inquiry-Based Module Reports

In the truest sense, science is apprenticeship-style learning by which the student learns from established scientists, often working with them on research projects and learning their "trade" while involved in authentic research investigations. Students locate and read numerous research articles, critiquing them so as to learn the elements of scientific research and communication.

The inquiry-based modules included in this lab manual are designed to teach students about scientific research by having them actually conduct research themselves. The inquiry-based modules not only contain information regarding scientific subject matter related to specific research topics, but also illustrate the scientific process: how science research is conducted and reported. They are designed to engage students in conducting their own research, stimulate higher-order cognitive thinking, and ask students to extend the research being reported.

The following guidelines are suggested for using these inquiry reports:

1. Have students critique the research report in the inquiry module (see sample rubric for peer reviews provided in Appendix 3).

2. Have students come up with their own way of extending the research reported in the module, coming up with their own experimental design (a detailed explanation of the process in the next section and a sample rubric provided later).

3. Have students conduct a background literature search to discover what is known about the research topic they are focusing on.

4. Have students describe their expected results as well as graph these expected results (if possible, also conduct a short pilot study).

5. Have students run their experiment, collect and analyze data, and graph their data.

6. Have students discuss their results and draw conclusions.

7. Finally, have students point to future research on the research topic, discussing ways of extending their research findings.

Creating a Research Proposal

One of the most challenging aspects of scientific research is synthesizing past work, current findings, and new hypotheses into research proposals for future investigation. Research proposals require the careful and thoughtful construction of the conceptual framework, the specific questions and hypotheses of the proposed research, a detailed experimental design and methods that will be used, the projected analysis of the data, and the significance of the proposed research.

The introduction provides background information and places the proposed study within a conceptual framework. Previous research is summarized, leading to the question the current research project proposes to answer. For purposes of clarity, explicit hypothesis/hypotheses are given, with a general statement regarding the main experimental design. (Think of an hourglass, leading from a wide section to a narrow neck. The wide section is the larger conceptual framework or context of the study; the narrow neck is the specific research focus or research question. Later, in the "significance of the proposed research" section, you will again lead to a large section, moving from the specific results of your study to relating these to the larger conceptual framework again.)

The methods section provides more details regarding the experimental design and methodologies you will use. This section is where specific predictions, and the underlying rationales justifying those predictions, are discussed. Background information can be given regarding the particular study site, focal organism, or process being investigated. How you will conduct the study—that is, the specific methods you will employ—are discussed in this section.

The results section for a proposal describes the type of analysis you will use: What type of statistics will be used to analyze the data?

The final section of a research proposal discusses anticipated results and the significance of the research findings. This section is where you expand back into the larger conceptual framework of the study. How will your specific results lead to advances in the field of study? Does your study address a topic that lacks quantitative data, or are their untested assumptions? This section should be pretty clear-cut if you did a thorough enough background investigation. The background review should have provided a clear path to an area within the field of study that requires more research.

Sample rubric for experimental design:

1. What is the research question you will investigate?

2. What is the significance of this question? How will the answer you discover contribute to and advance our understanding?

3. What is your hypothesis and prediction(s) (i.e., what do you *expect* to see *if* your hypothesis is accurate)? (Use the *If . . . , then . . .* format)

4. What is your experimental design? For example, control versus experimental group? Sample size in each group? What you will measure or take data on, your independent and dependent variables, and any statistical tests you will use.

CONTROL GROUP	EXPERIMENTAL GROUP

5. Describe the methods you will use so that someone could read your description and repeat your experiment exactly:

_____

_____

_____

_____

_____

_____

_____

_____

_____

_____

_____

_____

_____

6. Create a data table of the data you *expect* to see if your hypothesis is accurate.

7. Create the *data-recording tables* you will use when you do your experiment; give a copy to your instructor.

8. Create a graph of your expected data and turn it in to instructor.

9. Make sure your equipment is assembled and ready and that your experiment will actually work! Do a *pilot study* run-through with a sample size of 1 in each group.

When you run your experiment, remember that you must be able to organize and graph your data, present your findings to the class, and interpret your results in terms of whether or not your hypothesis is supported or rejected. Be sure to explain your results.

Resources

Day, R. A. 1988. *How to Write and Publish a Scientific Paper* (3rd ed.). ISI Press, NY.

Friedland, A. J., and C. L. Folt. 2000. *Writing Successful Science Proposals.* Yale University Press, New Haven and London.

National Science Foundation. *Grant Proposal Guide.* NSF Web site.

Genetics Problems

You may find it helpful to draw your own Punnett squares on a separate sheet of paper for the following problems.

Monohybrid Problems with Complete Dominance

1. In mice, black fur (B) is dominant over brown fur (b). Breeding a brown mouse and a homozygous black mouse produces all black offspring.

    a. What is the genotype of the *gametes* produced by the brown-furred parent? _____

    b. What genotype is the brown-furred parent? _____

    c. What genotype is the black-furred parent? _____

    d. What genotype is the black-furred offspring? _____

    e. If two F_1 mice are bred with one another, what phenotype will the F_2 offspring be, and in what proportion?

        phenotype _____

        proportion _____

2. The presence of horns on Hereford cattle is controlled by a single gene. The hornless (H) condition is dominant over the horned (h) condition. A hornless cow was crossed repeatedly with the same horned bull. The following results were obtained in the F_1 offspring:

    8 hornless cattle

    7 horned cattle

    What are the parents' genotypes?

        cow _____

        bull _____

3. In fruit flies, red eyes (R) are dominant over purple eyes (r). Two red-eyed fruit flies were crossed, producing the following offspring:

    76 red-eyed flies

    24 purple-eyed flies

    a. What is the approximate ratio of red-eyed to purple-eyed flies? _____

    b. Based on your experience with previous problems, what two genotypes give rise to this ratio? _____

    c. What are the parents' genotypes? _____

    d. What is the genotypic ratio of the F_1 offspring? _____

    e. What is the phenotypic ratio of the F_1 offspring? _____

Monohybrid Problems with Incomplete Dominance

4. Petunia flower color is governed by two alleles, but neither allele is truly dominant over the other. Petunias with the genotype R^1R^1 are red-flowered, those that are heterozygous (R^1R^2) are pink, and those with the R^2R^2 genotype are white. This is an example of **incomplete dominance.** (Note that superscripts are used rather than upper- and lowercase letters to describe the alleles.)

    a. If a white-flowered plant is crossed with a red-flowered petunia, what is the genotypic ratio of the F_1 offspring? _____

    b. What is the phenotypic ratio of the F_1 offspring? _____

    c. If two of the F_1 offspring are crossed, what phenotypes will appear in the F_2 generation? _____

    d. What will be the genotypic ratio in the F_2 generation? _____

Monohybrid Problems Illustrating Codominance

5. Another type of monohybrid inheritance involves the expression of *both* phenotypes in the heterozygous situation. This is called **codominance.**

One well-known example of codominance occurs in the coat color of Shorthorn cattle. Those with reddish-gray roan coats are heterozygous (RR'), and result from a mating between a red (RR) Shorthorn and one that's white ($R'R'$). Roan cattle don't have roan-colored hairs, as would be expected with incomplete dominance, but rather appear roan as a result of having both red *and* white hairs. Thus, the roan coloration is not a consequence of pigments blending in each hair. Because the R and R' alleles are *both* fully expressed in the heterozygote, they are codominant.

a. If a roan Shorthorn cow is mated with a white bull, what will be the genotypic and phenotypic ratios in the F_1 generation?

genotypic ratio _____

phenotypic ratio _____

b. List the parental genotypes of crosses that could produce at least some

white offspring _____

roan offspring _____

Monohybrid, Sex-linked Problems

6. In humans, as well as in many other animals, sex is determined by special sex chromosomes. An individual containing two X chromosomes is a female, while an individual possessing an X and a Y chromosome is a male. (Rare exceptions of XY females and XX males have recently been discovered.)

I am a male/female (circle one).

a. What sex chromosomes do you have? _____

b. In terms of sex chromosomes, what type of gametes (ova) does a female produce? _____

c. What are the possible sex chromosomes in a male's sperm cells? _____

d. Which parent's gametes will determine the sex of the offspring? _____

7. The sex chromosomes bear alleles for traits, just like the other chromosomes in our bodies. Genes that occur on the sex chromosomes are said to be sex-linked. More specifically, the genes present on the X chromosome are said to be X-linked. Many more genes are present on the X chromosome than are found on the Y chromosome. Nonetheless, those genes found on the Y chromosome are said to be Y-linked.

The Y chromosome is smaller than its homologue, the X chromosome. Consequently, most of the loci present on the X chromosome are absent on the Y chromosome.

In humans, color vision is X-linked; the gene for color vision is located on the X chromosome but is absent from the Y chromosome.

Normal color vision (X^N) is dominant over color blindness (X^n). Suppose a color-blind man fathers the children of a woman with the genotype $X^N X^N$.

a. What genotype is the father? _____

b. What proportion of daughters will be color-blind? _____

c. What proportion of sons will be color-blind? _____

8. One daughter from the preceding problem marries a color-blind man.

a. What proportion of their sons will be color-blind? (Another way to think of this is to ask, What are the *chances* that their sons will be color-blind?) _____

b. Explain how a color-blind daughter might result from this couple.

_____

_____

Dihybrid Problems

Recall that pigmented eyes (P) are dominant to nonpigmented (p), and dimpled chins (D) are dominant to nondimpled chins (d).

9. A pigment-eyed, dimple-chinned man marries a blue-eyed woman without a dimpled chin. Their first-born child is blue-eyed and has a dimpled chin.

  a. What are the possible genotypes of the father? _____

  b. What genotype is the mother? _____

  c. What alleles may have been carried by the father's sperm? _____

10. Suppose a dimple-chinned, blue-eyed man whose father lacked a dimple marries a woman who is homozygous recessive for both traits.

  a. What is the expected genotypic ratio of children produced in this marriage? _____

  b. What is the expected phenotypic ratio? _____

11. In his original work on the genetics of garden peas, Mendel found that yellow seed color (YY, Yy) is dominant over green seeds (yy) and that round seed shape (RR, Rr) is dominant over shrunken seeds (rr). Mendel crossed pure-breeding (homozygous) yellow, round-seeded plants with green, shrunken-seeded plants.

  a. What will be the genotype and phenotype of the F_1 produced from such a cross?

    genotype _____

    phenotype _____

  b. If the F_1 plants are crossed, what will be the expected phenotypic ratio of the F_2 generation? _____

Multiple Alleles

12. The major blood groups in humans are determined by **multiple alleles;** that is there are *more than* two possible alleles, any one of which can occupy a locus.

    In this ABO blood group system, a single gene can exist in any of three allelic forms: I^A, I^B, or i. The alleles A and B code for production of antigen A and antigen B (two proteins) on the surface of red blood cells. Alleles A and B are codominant, while allele i is recessive.

    Four blood groups (phenotypes) are possible from combinations of these alleles (Table 1).

TABLE 1 The ABO Blood Groups			
Blood Type	**Anitgens Present**	**Antibody Present**	**Genotype**
O	Neither A nor B	A and B	ii
A	A	B	$I^A I^A$ or $I^A i$
B	B	A	$I^B I^B$ or $I^B i$
AB	AB	Neither A nor B	$I^A I^B$

  a. Is it possible for a child with blood type O to be produced by two AB parents? _____ (yes or no) Explain

  _____

  _____

  _____

  b. In a case of disputed paternity, the child is type O, the mother type A. Could an individual of the following blood types be the father? _____ Explain each possibility.

    O _____

    A _____

    B _____

    AB _____

Chi-Square Analysis

13. In fruit flies, red eyes (R) are dominant over white eyes (r). A student performs a cross between a heterozygous red-eyed fly and a white-eyed fly. The student counts the offspring and finds 65 red-eyed flies and 49 white-eyed flies.

   a. What is the expected phenotypic ratio of this cross? _____

   b. Using a χ^2 test, determine whether the deviation between the observed and the expected is the result of chance.

   $\chi^2 =$ _____

   c. Conclusion

   _____

   _____

14. In fruit flies, gray body (G) is dominant over ebony body (g).

   a. A red-eyed, gray-bodied fly known to be heterozygous for both traits is mated with a white-eyed fly that is heterozygous for body color.

   What is the expected phenotypic ratio for this mating? _____

   b. The observed offspring consist of 15 white-eyed, ebony-bodied flies; 31 white-eyed, gray-bodied flies; 12 red-eyed, ebony-bodied flies; and 38 red-eyed, gray-bodied flies.

   What is the χ^2 value for this cross? _____

   c. Is it likely that the observed results "fit" the expected values? _____

Terms of Orientation in and Around the Animal Body

Body Shapes

Symmetry. The body can be divided into almost identical halves.

Asymmetry. The body cannot be divided into almost identical halves (for example, many sponges).

Radial symmetry. The body is shaped like a cylinder (for example, sea anemone) or wheel (for example, sea star).

Bilateral symmetry. The body is shaped like ours in that it can be divided into halves by only one symmetrical plane (midsagittal).

Directions in the Body

Dorsal. At or toward the back surface of the body.

Ventral. At or toward the belly surface of the body.

Anterior. At or toward the head of the body—ventral surface of humans.

Posterior. At or toward the tail or rear end of the body—dorsal surface of humans.

Medial. At or near the midline of a body. The prefix *mid-* is often used in combination with other terms (for example, midventral).

Lateral. Away from the midline of a body.

Superior. Over or placed above some point of reference—toward the head of humans.

Inferior. Under or placed below some point of reference—away from the head of humans.

Proximal. Close to some point of reference or close to a point of attachment of an appendage to the trunk of the body.

Distal. Away from some point of reference or away from a point of attachment of an appendage to the trunk of the body.

Longitudinal. Parallel to the midline of a body.

Axis. An imaginary line around which a body or structure can rotate. The midline or *longitudinal axis* is the central axis of a symmetrical body or structure.

Axial. Placed at or along an axis.

Radial. Arranged symmetrically around an axis like the spokes of a wheel.

Planes of the Body

Sagittal. Passes vertically to the ground and divides the body into right and left sides. The *midsagittal* or *median plane* passes through the longitudinal axis and divides the body into right and left halves.

Frontal. Passes at right angles to the sagittal plane and divides the body into dorsal and ventral parts.

Transverse. Passes from side to side at right angles to both the sagittal and frontal planes and divides the body into anterior and posterior parts—superior and inferior parts of humans. This plane of section is often referred to as a cross section.

Illustration References

Abramoff, P., and R. G. Thomson. 1982. *Laboratory Outlines in Biology III.* New York: W. H. Freeman.

Boolootian, R. A., and K. A. Stiles Trust. 1981. *College Zoology.* Tenth Edition. New York: Macmillan.

Case, C. L., and T. R. Johnson. 1984. *Experiments in Microbiology.* Menlo Park, California: Benjamin/Cummings.

Fowler, I. 1984. *Human Anatomy.* Belmont, California: Wadsworth.

Gilbert, S. G. 1966. *Pictorial Anatomy of the Fetal Pig.* Second Edition. Seattle, Washington: University of Washington Press. .

Glase, J. C., et al. 1975. *Investigative Biology.* Ithaca, New York.

Hickman, C. P. 1961. *Integrated Principles of Zoology.* Second Edition. St. Louis, Missouri: C. V. Mosby.

Hickman, C. P., et al. 1978. *Biology of Animals.* Second Edition. St. Louis, Missouri: C. V. Mosby.

Jensen, W. A., et al. 1979. *Biology.* Belmont, California: Wadsworth.

Kessel, R. G., and R. H. Kardon. 1979. *Tissues and Organs.* New York: W. H. Freeman.

Kessel, R. G., and C. Y. Shih. 1974. *Scanning Electron Microscopy in Biology.* New York: Springer-Verlag.

Lytle, C. F., and J. E. Wodsedalek, 1984. *General Zoology Laboratory Guide.* Complete Version. Ninth Edition. Dubuque, Iowa: Wm. C. Brown.

Patten, B. M. 1951. *American Scientist* 39: 225–243.

Scagel, R. F., et al. 1982. *Nanvascular Plants.* Belmont, California: Wadsworth.

Sheetz, M., et al. 1976. *The Journal of Cell Biology* 70:193.

Shih, C. Y., and R. G. Kessel. 1982. *Living Images.* Boston: Science Books International/Jones and Bartlett Publishers.

Stanier, R., et al. 1986. *The Microbial World.* Fifth Edition. Englewood Cliffs, New Jersey: Prentice-Hall.

Starr, C., and R. Taggart. 1984. *Biology.* Third Edition. Belmont, California: Wadsworth.

Starr, C., and R. Taggart. 1987. *Biology.* Fourth Edition. Belmont, California: Wadsworth.

Starr, C., and R. Taggart. 1989. *Biology.* Fifth Edition. Belmont, California: Wadsworth.

Starr, C., and R. Taggart. 2001. *Biology: The Unity and Diversity of Life.* Ninth Edition. Belmont, California: Wadsworth.

Starr, C. 1991. *Biology: Concepts & Applications.* Belmont, California: Wadsworth.

Starr, C. 2000. *Biology: Concepts and Applications.* Fourth Edition. Belmont, California: Wadsworth.

Steucek, G. L., et al. 1985. *American Biology Teacher* 471: 96–99.

Storer, T., et al. 1979. *General Zoology.* New York: McGraw-Hill.

Villee, C. A., et al. 1973. *General Zoology.* Fourth Edition. Philadelphia: W. B. Saunders.

Weller, H., and R. Wiley. 1985. *Basic Human Physiology.* Boston: PWS Publishers.

Wischnitzer, S. 1979. *Atlas and Dissection Guide for Comparative Anatomy.* Third Edition. New York: W. H. Freeman.

Wolfe, S. L. 1985. *Cell Ultrastructure.* Belmont, California: Wadsworth.